St. Louis Community College

Library

5801 Wilson Avenue
St. Louis, Missouri 63110

Distributed
Telecommunications Networks
VIA SATELLITES AND PACKET SWITCHING

Distributed

Telecommunications Networks

VIA SATELLITES AND PACKET SWITCHING

Roy D. Rosner

LIFETIME LEARNING PUBLICATIONS

Belmont, California

A division of Wadsworth, Inc.
London, Singapore, Sydney, Toronto, Mexico City

Jacket and text designer: Rick Chafian
Editor: Susan Weisberg
Illustrator: John Foster
Composition: Syntax International

Printed in the United States of America

1 2 3 4 5 6 7 8 9 10—86 85 84 83 82

Library of Congress Cataloging in Publication Data
Rosner, Roy D.
    Distributed telecommunications networks via
satellites and packet switching.
    Includes bibliographies and index.
    1. Telecommunication systems. 2. Packet
switching (Data transmission) 3. Artificial
satellites in telecommunication. I. Title.
TK5105.R66 1982   621.38′0413   82-4675
ISBN 0-534-97933-5

To Shirley and Harry,
Rose and Irving

Contents

Preface

WHAT THIS BOOK DOES

As the communications marketplace has evolved from the monopolies of the regulated common carriers to the competition of a wide range of high-technology companies, it has become increasingly important for the telecommunications user to be technically knowledgable about the available services and facilities. Modern distributed telecommunications networks, which are required to meet the information transfer needs of even very modestly sized organizations, must be synthesized from a great array of components, devices, facilities, and services. Not to draw from the total marketplace can result in user end costs often many times as great as the "optimum" solution.

This book is intended to give both technical and managerial insight into the modern distributed networks needed to support the telecommunications requirements of this information age. Particular emphasis is placed on the movement of information within the machines of the user—that is, the data communications. We also study the various techniques available for integrating voice and nonvoice telecommunications services into common networks. We give considerable attention to communications satellites and packet switching, two of the most exciting and powerful new communications technologies for meeting user needs in as flexible and cost effective a manner possible.

INTENDED AUDIENCE

This book is intended as both a learning tool and a reference for systems analysts, computer scientists, engineers, teleprocessing system planners, and managers of organizations that supply communications services to the public at large or to their own organizations. It can also aid systems analysts, communications staff officers, and managers of network users in evaluating services to be obtained. In other words, it is meant to assist anyone who needs a working knowledge of the design considerations, alternatives, and techniques for meeting present and future telecommunications requirements.

The book is a direct result of my own involvement in large-scale telecommunications projects for the U.S. Government. The material has evolved from a number of advanced education seminars and short courses. The information to a large extent supplements and extends many of the ideas presented in my earlier book, *Packet Switching: Tomorrow's Communications Today*. If you have read *Packet Switching*, you will find some of the material in the current book, particularly in Chapters 4, 5 and 6, a useful review and summary of the key functional and operational aspects of packet switching. In addition, certain ideas that are only briefly introduced in the earlier book, such as analysis of random network distributions or network security, are treated much more thoroughly here.

If you have not read *Packet Switching*, you may find yourself with some unanswered questions about the details of packet switched networks. The readings listed at the end of Chapters 4, 5, and 6, as well as the earlier book, will be useful in this case.

STRUCTURE AND ORGANIZATION OF THE BOOK

Chapter 1 begins our examination of modern distributed telecommunications networks with a general introduction to the concept of shared resource networks and the characteristics of different types of network users. Chapter 2 elaborates on some of the cost and performance features that make communications satellites a major resource in widely dispersed networks encompassing large geographic areas. In Chapter 3 we present a technical comparison of various techniques for sharing the resources of a communications network.

In Chapter 4 we describe packet switching, first in terms of its current application to data networks, and then in terms of its possible applications to voice communications. Chapters 5 and 6 discuss the ability to establish packet communication networks without discrete packet switches, using random broadcast techniques over radio, satellite, and cable media. We present a wide range of algorithms, which achieve utilization efficiencies of the transmission medium ranging from 18% to nearly 100%. Chapter 7 extends these ideas further, concentrating on terrestrial systems, particularly those applicable to localized geographic areas.

In Chapters 8 and 9 the very important technology of local area networks is described in detail. These networks, using many of the principles and techniques developed in earlier chapters, are fundamental in moving data among a large number of users in a local geographic area—within a building, an industrial complex, a campus, or a town or city. Most important, local area networks can significantly affect the overall costs of communications functions.

Shared, distributed networks permit each user to easily access the other users of the network. Thus it is increasingly important for users to protect their private information from intentional or inadvertant disclosure through the network facilities. Chapter 10 deals with the methods available to protect information in a common user, shared resource network.

Chapter 11 presents a number of different approaches to integrating voice and data users into a common network. Many of the local networking techniques provide such capabilities; extending these to nationwide networks is more of a technical challenge, which may be carried out more effectively by a small organization than by a major common carrier, at least in the short term.

Chapter 12 applies all the concepts of the book in a specific case study of a large, nationwide distributed network, utilizing a large amount of satellite connectivity. In Chapter 13 an approximate random network approach to essentially the same network design problem is derived, and the results are compared to the more comprehensive techniques of Chapter 12.

Chapter 14 concludes our study of modern distributed networks by introducing the range of available and projected common carrier and commercial services that provide the resources needed to implement the concepts developed throughout the book.

This book is intended to be largely self-contained. An annotated list of carefully chosen supplementary readings at the end of each chapter can lead interested readers beyond the principles of distributed networks and into the specifics of hardware design or software coding. However, the emphasis of this book is on the principles of distributed network operation and the relative advantages of the various techniques. Equally important is the understanding that modern cost-effective telecommunications networks result from the synthesis of many different components, each supplying a particular part of the overall requirements.

ACKNOWLEDGMENTS

Much of my personal involvement with the technology of distributed communications networks took place during my tenure at the Defense Communications Engineering Center in Reston, Virginia. I would like to gratefully acknowledge the far-sighted leadership and support of the people involved with the development of many of the ideas presented here. In particular, Dr. Harry VanTrees and Dr. Irwin Lebow, during their respective terms as the Chief Scientist of the Defense Communications Agency, were instrumental in establishing and encouraging the development of advanced concepts to meet tomorrow's needs with the most flexible and effective technology possible. I would also like to thank Professor Norman Abramson, who provided me with much of the insight and understanding of broadcast and random access networks, and the use of small satellite earth stations in large distributed networks. Finally, I would like to thank my present management at the U.S. General Services Administration for the challenging assignment of applying the technology discussed in this book to the very real problem of supplying low-cost, highly reliable telecommunications service to the government of the United States.

A very special thank you is in order for Judith, Stuart, Matthew, and Rachel, who did not see very much of me while this project was in progress. The support of each of you has been essential to its success.

Roy Daniel Rosner

April 1982

Distributed
Telecommunications Networks
VIA SATELLITES AND PACKET SWITCHING

1

Resource Sharing in Distributed Communications Networks

THIS CHAPTER:

will introduce the concepts of distributed data communications networks.

will show the relationship between user demand and the effective sharing of available resources.

will show how intelligent user end devices, packetized communications, and satellite transmission can all combine to produce state-of-the-art telecommunications capabilities.

Dick Tracy has had one for more than fifty years, so why shouldn't you have one by 1985? What is being referred to, of course, is the "wrist-watch" communicator, which provides instant electronic communication from wherever you are—in other words, the ultimate in distributed communications, the miniature personal terminal. This idea is neither science fiction, nor a prophecy of future technology. The current rapid development of numerous communications techniques will make such capabilities as personal portable terminals both feasible and economically viable. Two technological developments in particular—communications satellites and packet switching—are the largest components in achieving highly responsive, user-oriented, distributed telecommunications systems. As we progress through this book, we shall discover how the synergy of satellites and packet switching makes highly distributed networks a practical reality.

COMPUTER NETWORKS AND RESOURCE SHARING

Before we move further into specific applications, let us consider the general problem of resource sharing of communications and automated data processing (ADP) facilities. Figure 1-1 shows a generalized switched network, where a single

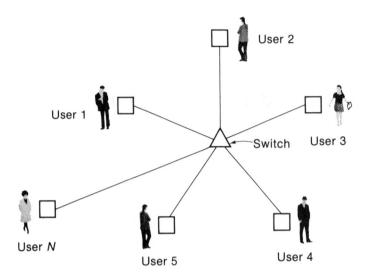

Figure 1-1. A generalized switched network.

switch permits an arbitrarily large community of N users to interchange information. The central element of this illustrative network is portrayed as a "switch," capable of interconnecting the user devices to each other by any one of a number of possible methods. In many applications, the switch may very well be a computer, shared among the community of users, and often functioning as an information resource as well as a switching device. In any case, this conceptual picture places each user at the end of a communications circuit, with the resources of each circuit devoted to the requirements of that specific user.

Figure 1-2 shows the same situation in a slightly more general way, indicating a centralized resource, M, and a number of individual terminal devices, each designated by T. However, an individual line or communications circuit is not provided between M and each T, as was the case in Figure 1-1. Instead, M and each T are connected by a shared medium, which could be a single, multidrop shared circuit, a shared terrestrial radio channel, or a common user broadcast satellite channel.

Allocating Resources through Protocols

Sharing of a common communications medium, regardless of its physical or electrical form, requires a set of rules, or a **protocol,** that governs the behavior of the group of users so that all users are able to gain access to the central resource. Implementation of operational protocols on a shared medium is complicated by the fact that the communication between the users and the central point is not symmetrical. In general, M is an intelligent, processor-based device, which can

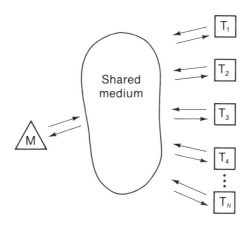

**Figure 1-2.   A communications network with users connected through
a shared medium.**

control the flow of information between itself and the large number of Ts. The
problem lies in the transmission of information from the various Ts to the central
point, M. The independent user terminals will undoubtedly attempt to send information over the common medium at random times, and since the Ts can
communicate with each other only via M, they have no direct way of knowing when
the common medium is already being used by other Ts. Consequently, some formalized structure is necessary so that the resources of the shared communications
medium can be fairly allocated among the large group of competing users.

Sharing of the common medium can be accomplished by various means.
Highly structured techniques, such as frequency-division multiplexing or time-
division multiplexing, achieve very disciplined operation of the shared medium
and extremely "fair" sharing of the total resources. In general, hardware and
equipment are somewhat complex, and resources tend to be wasted under normal
operational conditions. Less structured techniques, such as polling or multidrop
circuits, require less hardware and equipment and, in many cases, use the total
shared resources more efficiently. However, less structured techniques may produce temporary "unfairness" in the sharing process, and users may be delayed in
getting access to the shared medium. Essentially unstructured techniques, such as
the ALOHA channel and carrier sense multiple access (CSMA) techniques,
require little or no equipment complexity and produce surprisingly good performance. A key operational requirement for unstructured sharing of a common
medium is that each user be able (1) to assess the status of the shared medium, and
(2) to detect the success or failure of his own attempts to communicate over the
shared medium.

It is in the matter of assessing the status of communications that the synergy
between satellites, packet switching, and a large community of distributed communications users becomes most useful. Because of the universal coverage of the

information transmitted over the satellite channel, each user is able to very accurately assess the performance not only of his own communications attempts but also of the attempts of all other system users. By requiring users to communicate in packets—that is, bursts or blocks of information of predetermined maximum length—**packet switching** greatly facilitates fair and equitable sharing of the resource. These points will be developed in detail in subsequent chapters, as the technical characteristics of various sharing techniques are fully described.

The Intermittent Nature of Demand

The sharing of commonly available telecommunications resources is dependent on the intermittent nature of the users' demand. Periods of transmission are separated by periods of idleness, where the relative duration of the periods varies considerably, depending on the particular communications activity in progress. In voice communications, the idle periods are generally quite small relative to the active periods. Idle periods of a few seconds separate sentences and thoughts, and idle periods of fractions of seconds separate words and syllables. Averaged over a few minutes, or possibly a full hour, the total idle time compared to the total active time may be quite small. For many data communications activities, on the other hand, particularly for a remote terminal interacting with a computer, the idle time, compared to total active transmission time, may be quite large. On a percentage basis, generally less than 2% of capacity is used in data communications, whereas for voice communications typical occupancy is about 30% of channel capacity.

As an example of data communications use, let us consider an operational dialogue between an airlines reservations clerk and the reservations data base in the airline computer. A new customer requests flight information on a given date between a particular pair of cities. The clerk enters this information via the remote terminal, along with the appropriate command to call up flight and schedule information. This request would require approximately 20 typed characters, or about 160 data bits. Within a few seconds, the computer responds with the requested flight information, consisting of several thousand characters, possibly as many as 8000 bits of data. The clerk discusses this information with the customer, and, in the simplest case, a reservation request by the customer may result. The clerk would then request the customer's name, telephone number, address, or credit card information, and enter this data into the computer, again using bursts of 20, 30, or possibly as many as 100 characters at a time.

Once the customer identification information is entered, the computer confirms receipt, and the transaction is completed. For this simple case, the reservation clerk entered as many as 1000 data bits, and received from the computer approximately 10,000 data bits. With the required conversation between the clerk and the customer, thought, decision, and waiting time, the total transaction would typically take between three and five minutes (or 180 to 300 seconds). Thus, measured over the total time interval of the complete transaction, the reservation clerk sends an average of three or four bits per second, while receiving thirty to forty bits per second from the computer.

This customer-clerk scenario is typical of a wide range of terminal-to-computer activities associated with inquiry-response and interactive distributed processing systems, including interactive software development, banking and financial terminals, remote time-sharing terminals, retail and supermarket register and checkout terminals, and most other ADP-based systems that involve real-time interaction with a human operator. In all applications of this general type, the average entry rate of data from the human user to the computer via the remote terminal is just a few bits per second, while the average data rate in the reverse direction, from the computer to the remote user, is typically the same to possibly ten times as many bits per second.

Peak-to-Average Data Demand Ratios

Although *average* data transmission rates are quite low, the data rate when information is actually being transmitted has to be considerably larger in order to avoid excessive delays in sending and receiving data. The average data rate relates to the total amount of information that needs to be transferred, whereas the peak data rate relates to the "burstiness" and speed with which the information has to be delivered. Exchanges are characterized by both average and required peak data rates.

Table 1-1 presents the peak and average data characteristics for a number of different operational classes, in bits per second (b/s). The terminal-to-computer and computer-to-terminal cases are typical of the airline reservations scenario. Remote job entry terminals generally operate at higher volumes of data by grouping larger data processing jobs together for remote processing. Computer-to-computer transfers generally result from computer mainframes exchanging significant quantities of data. Such exchanges result when computers share their processing loads, or when large information distribution systems disseminate their

**Table 1-1. Low Duty-Cycle (bursty)
Communications Users**

	Average Rate	*Peak Rate*
Terminal-to-computer	1 b/s	100 b/s
Computer-to-terminal	10 b/s	10,000 b/s
Remote job entry	100 b/s	10,000 b/s
Computer-to-computer	10,000 b/s	1,000,000 b/s
PCM digitized speech	20,000 b/s	64,000 b/s

accumulated data. By comparison, the peak and average data rates readily attainable for pulse code modulation (PCM) of human speech exhibit a ratio of 3:1, much smaller than data communications cases, but significant nonetheless.

For all the classes of users shown, the peak data rates are 100 times greater than the average data rates, which means that a communications channel configured to meet the required peak data rate is likely to operate very inefficiently in terms of the average traffic load actually carried. The average required data rate, particularly where one end of the communications link involves a human "thought process," is very low, typically on the order of a few, or possibly as many as ten, bits per second. Stated in a different, slightly more mathematical way, the **interarrival time** between successive information bursts is long in comparison with the duration of each individual burst of information. In our airline reservations scenario, the 160-bit inquiry message would have a duration of 67 milliseconds (0.067 second) on a 2400-bits per second channel. This time is indeed very short compared to the few seconds or longer that would typically elapse until the next burst of data from the same user.

The relatively high cost of media with enough capacity to accommodate the required peak transmission rates has led to resource sharing of communications media. When media are shared among a large number of bursty users, the utilization tends to average out, yielding high average utilization as well as high peak capacity. Measured on an expanded time scale, almost all forms of communications have a degree of burstiness. Human speech, though it does have an element of continuity, is really made up of information bursts, separated by varying time intervals, as we have seen. Though not as dramatic as the 100 to 1 ratio typical of computer terminals, the peak-to-average ratio of human speech is at least 3 to 1; in other words, in normal conversation, at least two-thirds of the total time is consumed in idle periods.

THE TOTAL COMMUNICATIONS ENVIRONMENT

Communications networks are becoming increasingly multifunctional, whether they operate on a local, regional, national, or international basis. Local networks, particularly within a single office location or organization, are moving to increasingly complex and comprehensive combinations of computers and computer terminals, as well as to voice, graphical, and video communications devices. Real-time operation is often supplemented by textual message storage and retrieval systems, as well as systems that store voice information for either later delivery or voice-based retrieval systems.

The total communications environment thus combines a great variety of communications devices, a wide range of media, and a broad distribution of delay and throughput characteristics, as seen in the widely differing peak-to-average transmission ratios. Such an environment requires communications techniques that combine universality of coverage with a very high degree of adaptability. The property of universality is needed to tie the many disjoint user devices together, over both short and long distances, while the adaptability is needed to deal efficiently with the different traffic demands and device characteristics over a shared communications resource. Satellite channels provide universality of coverage, since within the view of a single synchronous satellite lie millions of square miles of the earth's surface with totally distance-independent communications capability. However, even though the cost is rapidly decreasing, satellite resources are relatively expensive and thus must be operated in a common user or shared mode. The protocols, or operational techniques, employed permit the sharing, as well as the adaptability needed to economically apply satellites to the total user communications environment. Supplemental facilities, such as wideband cables, that provide some of the local network concentration and information distribution can further extend resource sharing to earth stations as well as spaceborne satellite facilities. Thus, local networks of terrestrial facilities must operate in conjunction with satellite facilities to provide the appearance of a single, homogeneous communications medium, tying together a set of disjoint, nonhomogeneous user devices.

The Synergy of Network Components

The synergy between the satellite and terrestrial distributed networks enables them to operate successfully in the communications environment we have described. This synergy is facilitated by protocols based on the block-by-block transmission of information in bursts of previously specified maximum length. These blocks may be as small as a few hundred bits or as large as 10,000 bits in length, depending on many parameters of system operation. Proper sizing of the blocks involves knowledge of the user information characteristics, the speed and error characteristics of the transmission medium, the delay over the channel, and the processing required in the channel. Each block, or packet, represents a fully usable, self-contained entity, to be delivered from one location in the network to another. The synergy of satellite communications with large distributed networks is enhanced by the fact that packet-based techniques and protocols operate equally well in either satellite or terrestrial wideband channels. While specific details of the protocols must differ because of the delay over the satellite channel, the functions of the packet operation are essentially the same.

Thus, it becomes not only possible, but highly desirable, to achieve universal coverage of very large user populations by combining distributed local networks with the distance-independent operation of the satellite channel. This synergy permits users to access information from anywhere in the network with equal ease, regardless of the location of the users.

SUMMARY

1. The three components of large, modern communications systems are diverse user devices, broad bandwidth transmission media, particularly cable and satellite facilities, and the packet switched mode of operation.

2

Satellite Communications—Not a Cable in the Sky

THIS CHAPTER:

will discuss the technological and economic trends in satellite communications.

will outline the range of applications of satellite communications.

will introduce the relationship between satellites and distributed networks.

The words *LIVE VIA SATELLITE* flashed across millions of television screens nearly every day are probably the most vivid reminder of the success that the space program has had since the early 1960s. Satellite-based communications has added many dimensions to the media coverage of worldwide events, such as "live," real-time transmission of sports events and instantaneous news coverage. Not quite as visible to much of the world's population is the huge growth in high-quality international telephone service. It is particularly significant that the growth in service availability has not been limited to highly developed, highly industrialized nations; equally high quality service is available to all developing nations as well. However, no matter what its application, since its inception satellite-based communications has functionally been viewed as a giant "cable in the sky." By and large, satellite-based communications was carried out on a point-to-point basis, even though many pairs of points would be served simultaneously by a single satellite. It was only in the very late 1970s, with the increase of such applications as satellite-based cable television programming and demand assignment satellite-based networks, has it finally been recognized that the capability of satellites goes far beyond the cable in the sky.

COST, CAPACITY, AND DATA COMMUNICATIONS

Satellite Costs

Table 2-1 shows a number of the capacity and cost parameters associated with communications satellites since their first real commercial applications in 1965, projected through the late 1980s. These optimistic projections could be radically changed by even modest success of the United States space shuttle program, which not only would allow economical launch of larger and heavier payloads, but could ultimately lead to the practical application of repairable and reusable satellites. Organizations involved with the design and development of satellites have not yet begun to capitalize on the capabilities of the space shuttle for two reasons. First, the success and continued funding of the shuttle program were uncertain. Concern over this aspect of the shuttle has now been largely dispelled as a result of the highly successful initial test launches. Second, even assuming complete success of the shuttle, its launch capacity for the first five to seven years of its operation has been pretty well fully reserved for "conventional" satellites or evolutionary improvements of such designs. By conventional, we mean a satellite with a form factor and shape that could be launched by expendable rocket launch vehicles, rather than the shuttle. Thus, it will probably be 1987 or later before the launch of satellites that fully utilize the large weight and physical dimensions the space shuttle makes possible.

Figure 2-1 shows the historic trends of satellite communications costs and projects those trends to the late 1980s. Although the annual cost per voice channel is only one component of the total communications cost (that is, just the space-borne component), the impact of reduction in this cost will be impressive. Similar cost improvements in the other components of the total satellite communications system, particularly in earth station technology, will be experienced as well. Pres-

Table 2-1. Satellite Cost Estimates

INTELSAT Generation	Time Frame	Circuits	Design Lifetime	On-Orbit Cost	Annual Cost per Circuit
I	1965–1967	240	1.5 years	$8.2 million	$22,800
II	1967–1968	240	3 years	$8.1 million	$11,300
III	1968–1971	1200	5 years	$10.5 million	$1,800
IV	1972–1975	4000	7 years	$28 million	$1,000
IVA	1975–1980	6000	7 years	$26 million	$600
V	1981–1984	15,000	8 years	$34 million	$280
VA and beyond	Late 1980s	50,000+	10 years	$40 million or less	$80 or less

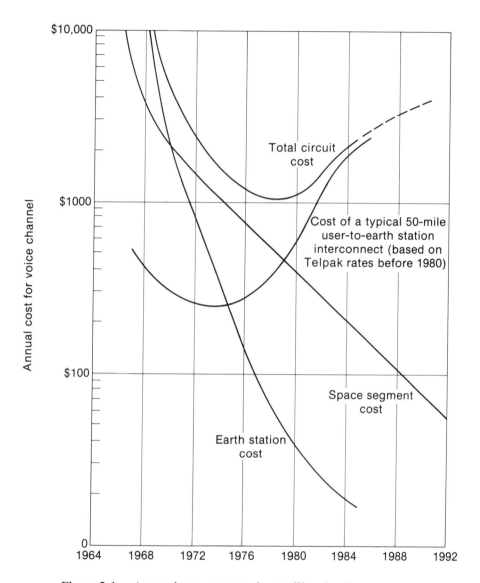

Figure 2-1.   Approximate cost trends, satellite circuit component.

ently, local distribution—that is, making the communications facilities accessible from remote user locations—is the major contributor to total system costs.

Other Cost Elements

Besides the cost of the satellite spacecraft, other elements of system cost must be considered. There is an allocated cost of the earth station, or a portion of an earth station if it is shared among many users. The cost of interfacing the user devices with the earth station may be inconsequential for a static user located close to the earth station, but it can be substantial for a mobile user or a user fairly far from his serving earth station. Finally, the user end devices, the voice terminals, the video displays, the keyboards, facsimile terminals, or any other user devices with specialized technical characteristics add to the cost of a communications system.

The cost approximations illustrated so far were based on a cost per equivalent voice channel of the space-borne portion of the satellite system. The derivation of individual voice channels over the satellite can be achieved in a number of ways, including time-division multiplexing (TDM) and frequency-division multiplexing (FDM), as well as more complex techniques. We shall describe the technical characteristics of each of these techniques in later chapters. At this point it is sufficient to note that TDM provides a fundamentally digital mode of operation of the satellite link, whereas FDM divides the satellite link into a large number of individual **analog channels.** Using an analog voice channel to carry digital information, such as computer data, requires **modems** (modulator-demodulators); it can then derive a data rate of approximately 9600 bits per second. However, as digital technology becomes more cost-effective with the operation of the satellite links based on time-division multiplexing, each voice channel equates to a digital bit rate of 64,000 bits per second. Some systems use lower rate techniques or voice-compression techniques, resulting in equivalent rates of 32,000 bits per second or even, in some instances, as low as 16,000 bits per second. In any case, a system based on TDM is highly capable of efficiently integrating high-speed digital data with voice and other communications services.

THE ULTIMATE DISTRIBUTED NETWORK

Chapter 1 introduced the ultimate in distributed networks, with everyone able to carry a portable communicator. Box 2-1 shows the derivation of the information transmission capacity available to each user based on the parameters of the INTELSAT V satellite. Box 2-1 assumes that a single satellite has all of its capacity devoted to the **UDN** (ultimate distributed network), and that the capacity of the satellite is shared equally by 220,000,000 people over a broad geographic area (say, the size of the United States). What this hypothetical derivation shows is that a single satellite provides enough capacity for each of the 220 million men, women, and children equipped with the wrist-band communicators to transmit nearly 40 pages of information over the system every day. In addition, by virtue of the satellite mode of operation, all of that transmitted information can be received by every other user within the UDN.

Before we all run out to buy our UDN terminals, we must realize that, at this time, our concept of the UDN is a little bit like the electric car that can go nonstop from New York to Los Angeles at 55 miles per hour average speed on $14 worth of electricity—unfortunately, the extension cord costs $2 million. While our UDN

INTELSAT V—equivalent capacity
17,000 voice circuits @ 64,000 bps/channel

total aggregate capacity $= 17 \times 64 \times 10^6$

$$= 1.088 \times 10^9 \text{ bps}$$
$$\times 3600 \text{ secs/hour}$$
$$\times 24 \text{ hours/day}$$
$$\overline{94 \times 10^{12} \text{ bits/day}}$$

$$\frac{94 \times 10^{12} \text{ bits/day}}{220 \times 10^6 \text{ persons}} = 427{,}000 \text{ bits per day per person}$$

$$\frac{427{,}000 \text{ bits per day per person}}{8 \text{ bits/letter} \times 5 \text{ letters/word} \times 275 \text{ words/page}} = 38.8$$

or approximately 39 pages per day for every person in the United States!

Box 2-1.

is possible from the point of view of satellite capacity, there are a few practical problems to be overcome, such as the earth station with the 90-foot diameter antenna.

Other considerations for a practical UDN include transmitter power, both on the **uplink** (from the earthbound user to the satellite) and on the **downlink** (from the satellite to the earthbound user); synchronization of the very high speed TDM digital links; separating that part of the transmitted data in which a particular user has an interest from the whole; and controlling the transmissions of each of the many users.

Though we will not be able to solve all the technical problems associated with the UDN in the next few years, as we proceed through the material in this book we will see that applying the principles of packet switching to the distributed satellite broadcast channel will provide many of the capabilities needed to move toward the realization of the UDN.

SATELLITE APPLICATIONS AND COMMUNICATIONS RESOURCE SHARING

The Development of Domestic Satellite Capacity

The broad application of satellites to many of the uses that evolved in the mid-1970s was not even envisioned as little as five years before the first U.S.

domestic satellites were launched in 1974. Satellites function most effectively and efficiently as a high-capacity trunking medium between two telephone switching offices. When the per-channel costs were very high, operational modes were chosen in order to make the limited capacity of the satellites available to the largest possible user community. In addition, early satellites were used for communication over very long distances and over regions of the earth difficult to span by other communications means. Consequently, satellites were used primarily for international and intercontinental communications. As the capacity of satellites grew, and the relative cost per channel rapidly shrank, satellites were increasingly used for less demanding applications.

Through the mid- and late 1960s, there was considerable debate about the need for national or domestic satellite communications. Leading the arguments against domestic satellite systems, the American Telephone and Telegraph Company (AT&T) insisted that domestic telephone users would not tolerate the service impairments (particularly the propagation delay) associated with satellite circuits compared to high-quality terrestrial communications facilities. This argument was based on the notion that users are more tolerant of service impairments in international telephone circuits. Although the circuits derived from satellite facilities are generally far superior to very long cable or radio circuits used for international calls, due to delay, they are generally less acceptable compared to terrestrial circuits. In addition, there seemed to be no anticipated demand for domestic services that could not be met with terrestrial systems. Despite AT&T opposition, the potential business opportunities of space-borne communications led to the "open skies policy." Established by a revolutionary decision of the U.S. Federal Communications Commission in 1972, this policy in effect permitted open entry of any fiscally responsible party into domestic satellite communications services. Plans for domestic systems were quickly announced by Western Union, Radio Corporation of America (RCA), COMSAT General Corporation, American Satellite Corporation, and others. To some extent, as AT&T had predicted, capacity of the domestic systems considerably exceeded demand. By 1975, with several domestic systems operational within the United States, as well as the Canadian ANIK system, which provided excellent U.S. coverage, a considerable amount of satellite capacity was available.

Finding Uses for Domestic Satellites

With a large amount of capacity available in the domestic systems, satellite owners began looking for innovative ways, beyond voice telephone traffic, to use that capacity. American Satellite Corporation leased transponder capacity on the Canadian system and on the Western Union Westar satellites in large increments and retailed it to end users in smaller increments, pioneering the concept of customer-premises earth stations—satellite earth stations, tailored to the needs of a particular customer, deployed at or near the physical location of the end users. The distributed nature of satellite operation, combined with customer-premises earth stations, opened the entire field of low-cost broadcast communications networks. Meanwhile, excess satellite capacity was made available to television and radio networks for program distribution and network broadcasting.

Application of satellite-based television distribution, coupled with the growth of local cable television for special programming, led to the single largest use of satellite capacity, and, in fact, to an effective "shortage" of total transponder capacity by the early 1980s. Television program distribution via satellite has also had the very positive effect of rapidly enhancing the state of the art in satellite earth stations, leading to the development of very low cost hardware capable of capturing the faint signals radiated from more than 22,300 miles above the earth.

By the late 1970s the Federal Communications Commission had contributed to the rapid increase in satellite system usage by radically simplifying the licensing procedures for earth stations, particularly receive-only stations needed for television distribution. Furthermore, between 1974 and 1981, the cost of a television receive-only earth station had declined from nearly $100,000 to under $10,000, and with further increases in satellite transmit power and improved receiver technology, broadband receive earth stations will continue to drop in cost. A COMSAT subsidiary is developing the capability to broadcast television programming directly to users' homes with an earth station a mere 30 inches in diameter and a projected cost of under $200.

All of this progress in distributed satellite broadcasting, both achieved and anticipated, has further broadened vision for satellite communications. Concepts and experiments of the early 1970s have become the practical reality of the early 1980s. Though total satellite capacity utilization still is, and probably will remain for many years, dominated by voice communications and television distribution, satellites are increasingly able to serve the needs of relatively low data rate information resource users.

SATELLITES AND DISTRIBUTED NETWORKS

As a result of the technological improvements and competitive activity in satellite communications, the availability of communications capacity has gone full circle. An initially very scarce and very expensive medium, domestic satellite capacity in the United States grew rapidly until total satellite capacity was in oversupply during the mid-1970s. As much of the capacity was converted to broadcast television use, however, satellite capacity is again severely limited. Looking forward to the mid- to late 1980s, it appears that technology will once again save the day, with the development of satellites that use much higher frequency bands (12/14 and 20/30 Ghz), and larger and more sophisticated transmission systems made possible by the size and weight launch possibilities of the space shuttle.

Until total available satellite capacity again balances total demand, the use of satellite facilities for distributed voice and data networking will require careful engineering to insure cost-effectiveness. A fairly large amount of total information will have to be aggregated at concentration points so that the satellite facilities can be heavily loaded at or near capacity during the busiest parts of the day. At the same time, the ability to apply the satellite resources on a demand-assigned basis allows the flexibility to accommodate demands whose geographic and occupancy patterns change rapidly.

The route to the "ultimate distributed network," then, is via an overall concept shown generically in Figure 2-2. Highly efficient, dynamic techniques will be used to tie many communications users together in a single local area. Local areas will be tied together in some cases with conventional communications facilities, and in other cases with some of the newer techniques, such as cellular radio, microwave distribution systems, or fiber optical systems. Over longer distances, satellite communications will tie the local distributions systems to each

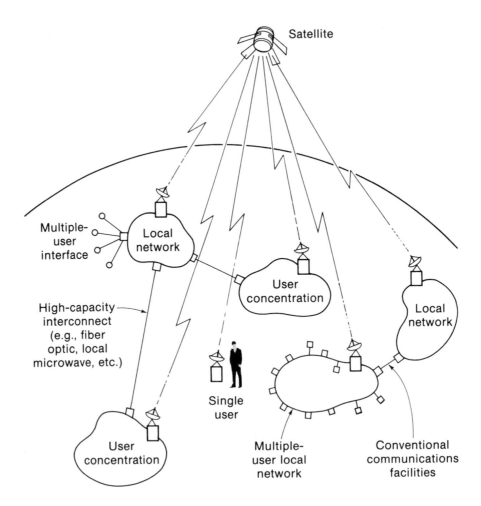

Figure 2-2. Generic distributed communications network.

other. By proper design of the interface protocols and operational procedures, end user-to-end user communications will be able to take place almost as if the users had direct access to a broadcast satellite channel. Thus, as the satellite capacity increases and satellite power requirements decrease, users will be able to change over to more direct satellite access. In addition, users in low-concentration areas—that is, areas remote from the high-capacity user concentrations—will gain direct access to all other users in the system more economically. We will roughly trace this probable evolution of distributed networks as we proceed through the book.

SUMMARY

1. Though satellite communications is still a scarce, expensive resource, technology is continually decreasing per-channel costs. However, in many cases local distribution, or local interconnection between a user and the nearest satellite service point (earth station), totally dominates the cost of the satellite service itself.

2. The inherent capacity of satellite systems is huge. In theory, at least, a single satellite could provide enough transmission resource to permit every person in the United States, for example, to transmit approximately 40 pages of unique information per day. Efficient use of this huge capacity will lead to universal access to large distributed networks, even using small portable terminals.

3. Satellite facilities will evolve to be just one portion of the overall structure of modern distributed telecommunications networks, providing in many cases the transmission resource between concentrations of users who are tied to high-capacity local communications networks.

SUGGESTED READING

BARGELLINI, PIER L. "Commercial U.S. Satellites." *IEEE Spectrum*, vol. 16, no. 10 (October 1979), pp. 30–37.

This survey of communications satellites looks at their effect on domestic networks, and the near- and long-term future of nationwide satellite communications. The paper also emphasizes the availability of relatively inexpensive, user-premises earth stations, as well as the ability of satellites to serve mobile users.

BOND, FREDERICK E., and ROSEN, PAUL. "Introduction and Overview." *IEEE Transactions on Communications*, vol. COM-27, no. 10 (October 1979), pp. 1377–1380.

The introduction to a special issue of the IEEE Transactions on Communications *on Satellite Communications, this article provides a brief summary of the state of the art of satellite communications. It also includes an excellent comprehensive bibliography and list of references.*

3

Multiple Access and Resource Sharing Techniques

THIS CHAPTER:

will contrast structured and unstructured approaches to
resource sharing.

will develop an understanding of several approaches to
communications facility sharing.

will introduce broadcast and narrowcast concepts of
resource sharing.

The statistical nature of the typical communications user together with the generally high peak-to-average ratio of individual user demand are the keys to high utilization of shared facilities. However, maximum utilization cannot be achieved without the use of relatively unstructured sharing techniques, which allow the maximum possible flexibility in assigning the total capacity of the communications medium. Let us begin by comparing structured and unstructured techniques of resource sharing.

STRUCTURED AND UNSTRUCTURED TECHNIQUES

The most common methods of sharing communications resources employ structured techniques. This simply means that capacity is assigned according to a set of rigid rules, which are often applied by means of a specific set of equipment or hardware. For example, in the ordinary switched telephone network, the total resources of the network are, in principle, available to all of the users, to be used in the network's attempt to complete a call between a given pair of user endpoints. The connection between two users is established by a sequence of steps, often employing physical electrical connections through the switching and transmission equipment, which provides the path over which the communication proceeds. Once assigned, the path between the users is effectively allocated on a full-time

19

basis to the particular user pair. Moreover, once a portion of the resources is so assigned, it is effectively removed from the total resource pool, and any capacity unused by the currently assigned users is inaccessible to any other users.

Unstructured techniques are based on the statistical nature of the users, with commonly available resources assigned in such a way that they are returned to the overall resource pool whenever the demands of the current user permit. The resources are managed to utilize idle periods in one user's activities to serve the needs of another. Such dynamic resource management results largely from applying computer processing to the control and operation of the communications transmission facilities.

Table 3-1 summarizes some of the key attributes of both structured and unstructured communications resource sharing techniques. These attributes are described in more detail below.

Typical Examples

The most widely applied communications resource sharing technique is **multiplexing.** In a structured approach—either frequency-division or time-division multiplexing—total capacity is normally divided into a number of subunits, with each subunit assigned to a particular user for an extended period of time. Typical examples of unstructured techniques include statistical multiplexors, polling multidrop data circuits, the class of channels operated under ALOHA algorithms, and carrier sense multiple access (CSMA) facilities. Operation of each of these techniques will be discussed further in subsequent chapters.

Table 3-1. Structured and Unstructured Communications Sharing Techniques

	Structured Approach	Unstructured Approach
Typical examples	Time-division multiplexor Frequency-division multiplexor	Statistical multiplexor Polling ALOHA channel Carrier sense multiple access
Basic operation	Deterministic	Statistical
Implementation	Hardware driven	Software driven
Capacity assignment	Fixed	Demand
Channel control	External information	Internal protocols
Maximum traffic throughput	100% of basic channel rate (all users active)	18% to 100% of basic channel rate
Average traffic throughput	10% or less typical	50% or more typical

Basic Operation

Structured techniques operate deterministically; that is, their operation remains fixed according to a predetermined, known pattern. Unstructured techniques are statistical in operation and attempt to respond to the unpredictable bursts of information transmission inherent to the usage by a particular user at any given time.

Implementation

By virtue of their fixed operational characteristics, structured techniques tend to be hardware driven, with the system capacity allocated by the physical interconnection of the various components. Unstructured techniques generally use computerlike processors to carry out the sharing function under software control, which provides both the adaptability and the flexibility to operate under dynamic conditions.

Capacity Assignment

A major difference between structured and unstructured techniques is in the flexibility of the system capacity assignment. Structured techniques tend to use fixed assignment. The capacity, once assigned to a particular user, is retained by that user for a period of time that is long relative to any one burst of information. Unstructured techniques assign capacity on a demand basis, responding to rapidly changing user information flows.

Channel Control

Structured resource sharing techniques are generally controlled by facilities and information that are external to the basic flow of information over the communications channels. In switched systems, for example, the signaling information to set up a connection precedes the actual flow of the users' information, generally in a completely different format from the users' information. In fixed multiplexors, the control of the facility is inherent either in the hardware itself or in special control signals (such as frequency standards or synchronizing frames) built into the information flows. Unstructured techniques are generally controlled by operational protocols. The control information needed to manage the assignment and flow of user information is inherent to the flow and is generally disseminated in the channel itself, using formats closely resembling those of the users' information.

Maximum Traffic Throughput

Under ideal conditions, structured techniques can easily achieve a maximum traffic throughput of 100% of the basic channel rate. However, the ideal conditions pertain only when every user who has capacity assigned is actually transmitting information. Statistically, such a condition has a very low probability of occurrence if most of the users do, in fact, exhibit relatively high peak-to-average transmission ratios. Unstructured techniques can achieve a maximum traffic throughput anywhere from 18% to 100% of the basic channel rate. The ability to achieve the higher throughputs is dependent upon more complexity in the channel assignment protocols and more rigid control of information flow in the channel.

Average Traffic Throughput

The average traffic throughput achieved by a structured assignment technique is generally 10% or less of the basic channel rate. If each individual user is assigned sufficient fixed capacity to meet his peak transmission rate, then the overall system throughput can be no higher than the average-to-peak ratio for all the users assigned to the system. For the unstructured system, capacity is dynamically allocated to meet the users' needs, generally only when useful information is being transmitted, leading to overall system efficiencies of 50% or more of the basic channel rate.

TECHNICAL DESCRIPTION OF RESOURCE SHARING TECHNIQUES

Figure 3-1 presents the situation in which a number of user terminals are attempting to share a single line (the common medium) to a central location, which might be a shared computer or the entry switch of a larger communications

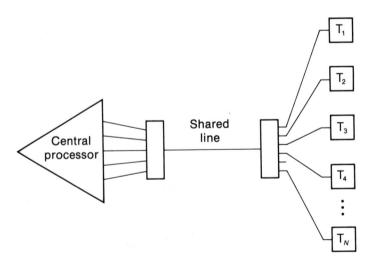

Figure 3-1.   A number of users attempting to share a single line to a central location.

network. The communications resource sharing techniques needed to accomplish the utilization of a single line vary considerably in complexity and operational sophistication. Simple multiplexors are the least complex and potentially the least efficient. Polling achieves many of the desirable attributes of the unstructured approaches, though it is highly controllable. Other techniques, such as random broadcast, are both unstructured and largely uncontrolled in the sense of a centralized resource manager. Several hypothetical cases will be used to illustrate the operation of many commonly used resource sharing techniques. In each case, the shared facility will be configured to accommodate 100 users, on a channel with a total bandwidth of 50 kilohertz (50,000 Hertz).

Frequency-Division Multiplexing (FDM)

If 100 users equally share the available 50,000 Hertz of bandwidth, each would have a gross bandwidth of 500 Hertz allocated to it. However, to prevent mutual interference on a frequency basis, about 40% of the available bandwidth would have to be utilized as guardband between adjacent channels, leaving a net bandwidth of about 300 Hertz. Using a multiple level modulation technique, the user channels would achieve about 2 bits per Hertz of bandwidth, resulting in a usable data rate of 600 bits per second (**BPS**) per user, or a total of 60,000 bits per second for the entire community of 100 users sharing the single line or common medium. Notice, however, that the only time the overall bit rate of 60,000 bits per second is actually achieved is in the unlikely event that all 100 users are transmitting data simultaneously. If fewer than all of the users are actually transmitting, the useful bit rate of the shared line falls below the apparent capacity. While a user is actually transmitting, the achieved bit rate of each individual user is only 600 bits per second. At this relatively low data rate, a typical user message of two or three typewritten lines (approximately 200 characters) will take about three seconds to transmit.

Although this particular example is somewhat contrived, and would not be practical to actually implement, it does serve to illustrate the principles involved. In practice, frequency-division multiplexing (**FDM**) techniques are employed for teleprinter applications where 16 or 24 low-speed teleprinters are frequency-division multiplexed onto a single 4000 Hertz voice channel. Such techniques, called **VFCT** for voice frequency carrier telegraphy, would provide, for example, up to 16 channels, each operating at 300 bits per second for a total maximum rate of 4800 bits per second.

Time-Division Multiplexing (TDM)

A more practical and more modern form of multiplexing utilizes time-division (**TDM**) techniques, where a single high-data-rate digital signal is time shared among the various users. With a good data modulation technique, a single 50,000 Hertz channel could achieve a total bit rate of 100,000 bits per second. If we divide the bit stream in time among 100 total users, each user will have an allocated total capacity of 1000 bits per second. Here again, maximum capacity is achieved only when all 100 users happen to be simultaneously active. While the 1000-bits per second data rate is nearly twice that achievable with frequency-division multiplexing, this low a data rate would still result in significant transmission delays for moderate length data messages.

The major shortcoming of both frequency- and time-division multiplexing is the fact that the capacity is preallocated to each of the potential users. As a result, capacity is wasted each time a potential user becomes inactive, and, at the same time, only a small part of the overall channel data rate is ever available to any individual user. Statistical multiplexing overcomes this disadvantage by assigning the capacity only to active users. Somewhat greater complexity is required at the terminal devices in the form of processing and buffering (temporary storage). In addition, some overhead is necessary in order to identify the information at each end of the circuit with the source and intended destination of each transmission.

Statistical Multiplexing Using Centralized Polling

A prevalent form of statistical sharing of a common medium is centralized **polling.** As shown in Figure 3-2, each user is accessed to a single common line. Associated with each user terminal is a **buffer,** which can temporarily store a block of data. When the user enters information to be sent to the central location, the information characters are held in the terminal's buffer. The buffered information is held until a poll request, sometimes called a transmit command or a write token, is received by the terminal from the controller. Upon receiving the poll request, the terminal is permitted by the system protocol, as implemented by the poll controller, to transmit information for a limited period of time—generally much less than a second, but usually long enough to empty a fully loaded buffer. After completing transmission, the poll request is then sent to the next terminal in turn. If a terminal has nothing to transmit, it will generally so indicate with a very short control message, and the poll request will move on to the next terminal.

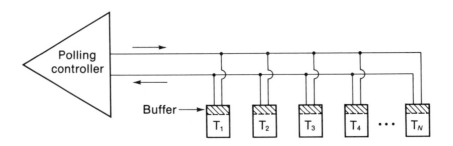

Figure 3-2.   Statistical sharing by means of centralized polling.

To understand how such a system might perform, we will again refer to the example of trying to serve 100 users with a single line of 50,000 Hertz available bandwidth. By use of a good, modern modulation technique, a gross capacity of 100,000 bits per second can be achieved on such a line. Each time a terminal is polled, the central controller has to wait a certain amount of time for the poll request to physically propagate along the transmission line to the distant terminal, and for the response to propagate back, before it knows if it can make a poll request to the next terminal. The physical propagation time depends upon the distance to the furthest possible terminal, which for this example we will assume to be about 100 miles. It will take about one millisecond for the signal to reach this distance, which, at 100,000 bits per second, translates to a transmit delay of about 100 bits.

Each time a poll request is made, the desired terminal has to be identified. Thus, an identification number, comprising at least 7 bits, is necessary for each of the 100 distinct terminals. In addition, some additional system overhead will be incurred, such as the control information that the terminal might send to indicate that it has no data to transmit on the current poll. Actual practice indicates that this additional overhead will typically be about 8 bits per poll. The total overhead in the channel is therefore 115 bits per poll (100 bits delay + 7 bits identification + 8 bits control). If each user is to be polled at least once per second (under idle channel conditions), then 11,500 bits per second of channel capacity is consumed by the channel management and overhead functions. This leaves approximately 88,500 bits per second of usable capacity, or an average capacity of 885 bits per second per user. This would be sufficient capacity, for example, for each of the 100 users to send a 100-character message about every 10 seconds.

The advantage of the polling techniques compared to the fixed multiplexing approaches described earlier lies in the fact that, when a terminal has its turn to transmit, it does so at the full line rate—in this hypothetical case, 100,000 bits per second. As a result, it would take less than one millisecond to transmit a 100-character message once the terminal received its transmission poll request. On a moderately loaded system, the transmit poll request should occur about once per second, on the average, resulting in much shorter delay times than in the cases of the hypothetical fixed multiplexors. Moreover, an inactive terminal does not consume or waste any of the transmission capacity.

There are many variations of the implementation of polling used in this example. Overhead can be reduced by making the terminals somewhat smarter and, for example, having them pass the poll request along to the next terminal rather than signal no traffic to the central controller. Other possibilities exist, and they have been used in operational systems. The key point is that polling is a prime example of a system which uses dynamic, rather than fixed, assignment of the capacity of the transmission line or shared medium. The available capacity is shared, following some formalized assignment procedures among the active members of the user community, and little or no capacity is wasted by the users that are currently inactive.

RESOURCE SHARING BY BROADCASTING AND NARROWCASTING

Broadcasting and Narrowcasting Techniques

As described above, polling provides for the communication of a large number of user terminals with a central controller, centralized processor, or switching facility. However, in many physical arrangements of actual facilities, it is possible for all devices in the system to hear the transmissions of all other devices, either directly, via relay from a terrestrial repeater station, or via a communications satellite. Such systems thereby become resource sharing via broadcasting.

The term **broadcasting** has traditionally meant the transmission of information from a single transmitter to a multitude of receivers. Radio and television broadcasting, of course, are the most common examples. In broadcast-based communications networks, whether for voice or data communications, a single centralized facility can converse with any of many possible destinations by means of self-identified information bursts or messages. Such messages are broadcast over the common medium and received by the proper addressee. Because the signals radiated from a communications satellite can be received equally over a very broad geographic area, such satellites make an ideal medium for highly distributed communications networks via broadcasting.

The inverse process—sending the data from any one of the dispersed users to the central location—can similarly be achieved by transmission of self-identified messages. Because of the convergence of many locations' traffic upon a single destination, we might refer to this form of transmission as **narrowcasting.** More generally, narrowcasting occurs when any one of many system users transmits information destined to one, or a small subset, of the total user community. When many dispersed users are narrowcasting messages over a shared channel type of system, there is a significantly high probability that some of those messages will be transmitted simultaneously, causing mutual interference.

Approaches to Shared Communications

Much of the remainder of this book will be devoted to the various approaches to capacity sharing and multiple-user networking using broadcasting and narrowcasting of self-identified blocks of communications traffic.

The techniques and principles of packet switched networks introduced in the early 1970s were based initially on the routing of self-identified message segments—packets—through terrestrial landline networks. Since that time, the concepts have been expanded to include broadcast and narrowcasting of data packets on shared media, including wire, coaxial cable, fiber optics, radio, and satellite channels. Packet communications, employing either computer-based switches or shared broadcast media, or a combination of both, have been developed to achieve high efficiency over a broad range of user requirements, for local as well as long-distance communications networks.

	Frequency-division multiplexing (FDM)	Time-division multiplexing (TDM)	Centralized polling	ALOHA
Average data rate (overall channel)	60,000 BPS (all users active)	100,000 BPS (all users active)	88,500 BPS	18,000 BPS
Peak data rate (single user)	600 BPS	1000 BPS	100,000 BPS	100,000 BPS
Number of users supported	100	100	100	9000
Advantages	Hardware driven Highly predictable "Fair" allocation of resources	Hardware driven Highly predictable "Fair" allocation of resources Matches data users' format well	Adds processing to hardware to achieve efficiency Permits access to full channel	Little hardware or processing Serves many users Gives access to full channel
Disadvantages	Static assignment is often not efficient Limited capacity for each user	Static assignment is often not efficient Limited capacity for each user	Additional complexity Some users may get more capacity than others	Low theoretical efficiency Poor control of channel operation

Figure 3-3. Comparison of structured and unstructured resource sharing techniques.

The simplest approach to broadcasting networks is the **ALOHA** protocol, which permits any user to transmit data at will, without regard for the current status of the channel. The average achievable data rate with this technique is only about 18% of the total channel capacity, but the channel is still capable of supporting the demands of a large number of users. For example, the 100,000-bits per second channel used in our earlier examples would be capable of an average rate of only 18,000 bits per second. However, each user would be able to transmit at a peak rate equal to the full channel rate of 100,000 bits per second. More important, such a channel would be able to support as many as 9000 users, compared to the 100 users supportable using the multiplexing and polling techniques.

Figure 3-3 is a tabular comparison of structured and unstructured resource sharing techniques. Note that, for the multiplexing cases, the high average data rates are achieved only in the unlikely case that all 100 users are simultaneously active and transmitting. The ALOHA technique achieves its apparent advantage from a form of **contention** that occupies no resources at all for an inactive user. Users compete—or "contend"—for system resources without recourse to a centralized controller or complex hardware channellization equipment. In addition, the technique requires no multiplexing equipment, no timing, and little or no formalized overhead structure. Its primary disadvantage—low theoretical efficiency—can be overcome by additional complexity in the operational protocols and terminal equipment, which will be demonstrated in subsequent discussions.

SUMMARY

1. The statistical nature of users permits effective use of shared communications facilities.

2. Resource sharing is accomplished by both structured and unstructured techniques.

3. Structured techniques, such as frequency-division multiplexing and time-division multiplexing, have poor overall efficiency because capacity is allocated inflexibly, regardless of whether users are active or not.

4. Unstructured techniques, such as statistical multiplexing and centralized polling, allow for the assignment of capacity on the basis of user requests and thus are more responsive and more flexible than fixed multiplexing.

5. Unstructured broadcast and narrowcast techniques, an outgrowth of packet switching technology in which self-contained, fully addressed packets of information are routed between users, offer the greatest promise of high-efficiency and flexibility in large, distributed communications networks.

SUGGESTED READING

DOLL, DIXON R. "Multiplexing and Concentration." *Proceedings of the IEEE*, vol. 60, no. 11 (November 1972), pp. 1313–1321.

This article is an excellent technical introduction to the various multiplexing and concentration techniques available to reduce the costs of data communications facilities. Several expressions are developed to express the queueing and delay times which user messages would experience in such systems.

SCHWARTZ, MISCHA. *Computer-Communication Network Design and Analysis.* Englewood Cliffs, N.J.: Prentice-Hall, 1977, ch. 12.

This chapter provides an introductory development of polling techniques and relates them to specific operational systems. Mathematical expressions and tables show the performance of various systems under different user loading conditions.

4

Distributed Networks and Packet Switching

THIS CHAPTER:

will introduce the history and general operational concepts of packet switched networks.

will show how packet switching meets the demands of short, bursty data communications users.

will describe the various ways that packet networks can operate to combine data, voice, and other modes of communications in a single common network.

Given the diversity of user applications and the myriad approaches to efficient resource sharing by application of distributed telecommunications networks, it is important to understand the concepts of packet communications. Packet switching has emerged as a telecommunications technique with unlimited potential for effective application. Public packet switching networks have been built or are planned in more than 20 countries around the world, and numerous private and experimental networks are currently using packet switching.

Because it permits communications resources to be used at utmost efficiency, packet switching can adapt to a wide range of user services and user demands. While it is most frequently used in computer and data communications, its effectiveness for voice, video, and other wideband telecommunications services has been demonstrated. As advanced data processing techniques improve the computer processors which form the heart of the packet switches, packet switching will have an even wider range of applications. More importantly, packet communications can ensue through broadcast operations without the need for actual packet switches. The concepts of packet broadcasting are central to many of the contemporary approaches to local area networks and the effective interconnection of diverse equipment and devices.

THE ORIGINS OF PACKET SWITCHING

Interestingly, the origins of packet switching are more strongly based in voice communications than in data communications, though packet switching is now more widely applied to data- and computer-based communications. Paul Baran and his associates at RAND Corporation in the early 1960s are generally credited with "inventing" packet switching. They were working to make military voice communications circuits safe from wiretapping and immune to interruption if part of a network was destroyed.

The idea started with the notion of breaking a voice conversation between two parties into short, separate pieces (packets). At each switch, the pieces of a call would be mixed with pieces of other calls and sent, piece by piece, over several different routes to the destination. Figure 4-1 shows the pieces of a particular call sent, packet by packet, over diverse routes through the network. For clarity, pieces of other calls, which would be interspersed with the packets shown, are omitted. Only at the destination would it be possible to collect all the pieces and, after reassembling them in proper order, make the voice intelligible. If the wires were tapped anywhere in the network, or if communications between microwave relay points were intercepted, all that would be heard would be the garble of dozens of interleaved bits and pieces of many conversations. Although these ideas were published in 1964, the technology was not really available to perform the complex processing, routing, and control functions required to implement this concept in a large-scale network.

In the meantime, the Advanced Research Projects Agency (ARPA) of the U.S. Department of Defense was supporting numerous large computer installations

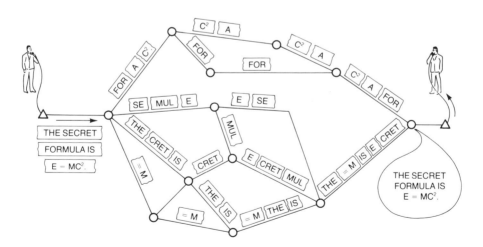

**Figure 4-1.   A packet switched network used for
voice communications.**

at various universities and laboratories throughout the United States that were doing basic and applied research. Because of differences in time zones, computer center workload, and specialized hardware and software, it seemed desirable to find an efficient way of sharing the resources. As no appropriate networking capabilities were available, ARPA embarked on developing them, and the result was the application of packet switching to data communications and the deployment of the **ARPANET** as the prototypical packet switched data network.

The "commercialization" of packet switching—largely through the efforts of Telenet (now GTE-Telenet); the Trans-Canada Telephone System (DATAPAC); the Postal, Telephone, and Telegraph Companies (PTTs) of France, Great Britain, and other countries; and numerous other network and equipment vendors—has evolved the ARPANET technology into a comprehensive, multilayered structure of networking systems, architectures, and protocols. By 1976 the Consultive Committee for International Telephone and Telegraph (**CCITT**) systems had adopted the widely accepted **X.25** standard for data user interface to a public packet switched network, allowing user equipment and software to be developed as commercial products to utilize packet switching services.

BASIC CONCEPTS

The concept of packet switching, using discrete switching elements in a distributed network, is based upon the ability of modern high-speed digital computers to act on transmitted information so as to divide the calls, messages, or transactions into pieces, called packets. Packets move around the network, from switching center to switching center, on a **hold-and-forward** basis; that is, each switch holds a copy of each packet in temporary storage until the switch is sure that it has been received properly by the next switch or by the destination user. This form of operation permits the network to achieve low overhead for short messages and eliminates the set-up time for calls that is required in conventional circuit switched telephone networks. Because all communications are broken down into similar component pieces, long messages and short messages can move through the network with a minimum of interference with each other. By moving the packets through the network in (nearly) real time, the switches can adapt their operation quickly in response to changing traffic patterns or failure of part of the network.

Within a given vendor's network, the internal network operation can be highly complex, with multiple paths, dynamic routing, multiple priorities, and many redundancy features. On the other hand, internal network operation can be quite simple, with the network performing nothing more than the functions of an asynchronous time-division multiplexor. The former approach is generally needed in order to achieve the advantages and reliability attributed to packet switched networks.

Significantly, the work at standardizing the user interface to packet switched networks did little to define how the internal operations of the network should proceed. Thus, many different implementations of packet switching have been developed, which, though they all provide standard user interfaces, are largely incompatible on a switch-to-switch level. As a result, the interoperation of different networks has to take place at a fairly high level of protocol, through interface **gateways,** as defined by the X.75 CCITT standard. The gateway approach in effect makes the entire network to which the originating subscriber is connected appear as a subscriber to another vendor's network, to which the destination subscriber is connected. The gateway is a process that transforms the internal message structure of the originating network into an input format accessible to the network that contains the destination subscriber.

Operation of a Hypothetical Network

Figure 4-2 illustrates a portion of an arbitrary packet switched network. Though there are seven switches (or **nodes**) in the network, we will focus on switches 1, 2, 3, and 4. User A is a subscriber attached to switch 1, and user B is a subscriber

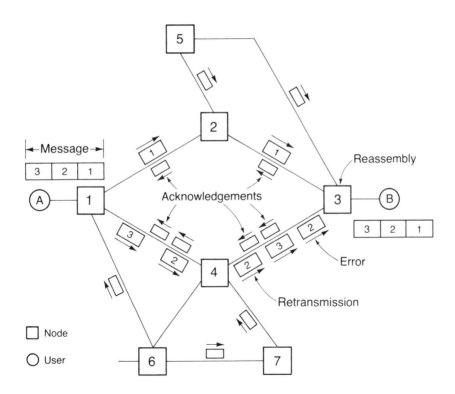

Figure 4-2.   Basic operation of a packet switched network: movement of a three-packet message from user A to user B.

attached to switch 3. We will trace the flow of a message from user A to user B, where the message is composed of three packets.

Modes of Operation. For short messages that can be fit into a single packet, the operation of any network is quite simple. The complete message, contained within a single packet, simply moves as a discrete entity from one user to the other, following a path through the network selected by the switching elements. In fact, the concept of message handling where the entire message is "required" to fit within a single packet is defined by a packet switching mode of operation known as **datagram.**

It is only when messages extend beyond the limit of a single packet that more complex and comprehensive data management protocols come into play. The ability of packet networks to handle multiple-packet messages requires the application of numerous checks and safeguards in the operational protocols to protect against various anomalies. What is known as **virtual circuit** operation of a packet switched network provides a number of steps and functions directly analogous to normal telephone operations. Virtual circuits have to be established through the network on the basis of a new **call request,** which, depending on the implementation, may or may not have any useful end-user data associated with it. Once the virtual circuit is established, packets flow on a continuous basis from user to user almost as if a fixed, permanent connection existed between the two user endpoints. Datagrams and virtual circuits will be discussed further later in this chapter.

Flow of the Message. Though Figure 4-2 focuses attention on the message flowing between users A and B, remember that other packets flowing between other users would be simultaneously moving throughout the network. The flow of the message is initiated by the transmission of packet 1 between user A and switch 1. Upon fully receiving the first packet, switch 1, following a set of routing rules, transmits packet 1 toward its destination by sending it via switch 2. In the meantime, packet 2 is moving from user A into switch 1. During this time, the conditions in the network change (for instance, a large amount of traffic from switch 5 arrives at switch 2), so the second packet of the message from A to B is routed via switch 4. The third packet of the message, arriving at switch 1 soon after the second packet, is similarly routed via switch 4.

After being received correctly by switch 4, the second packet is transmitted to the destination switch, switch 3, but during that transmission, an error occurs. Such errors can result from a burst of noise on the transmission path, static from lightning or switching equipment, momentary interruptions during automatic switchover from primary to backup equipment, and the like. When switch 3 receives packet 2, the error-detection mechanism is able to detect the error and requests a retransmission of packet 2. However, while this is occurring, packet 3 has been transmitted immediately behind the first (and errored) copy of packet 2. As a result, the second (correct) copy of packet 2 is received at switch 3 after packet 3. If the network is viewed from the perspective of switch 3, first packet 1 is received, then packet 3, and finally packet 2. If switch 3 delivered the packets to the destination (user B) in the same order they arrived at switch 3, user B would receive the packets in a different order than they entered the network.

Network-Introduced Problems

Note that Figure 4-2 shows a number of acknowledgements flowing on the various links in the network, in the opposite direction from the information packets. These acknowledgement packets are the key to the error-detection mechanism that is needed to insure the integrity and accuracy of the transmitted data. Any information packet that is properly received is immediately acknowledged back to the sender with one of these short acknowledgement packets. In this way, the sending switch knows that the information packet has been received properly by the next switch along the path toward the destination.

If an acknowledgement is not received within a certain period of time (known as the **timeout period),** the sending switch presumes that the packet was received in error and retransmits the packet. This presumption is necessary because it is quite possible that any transmitted packet could be so badly garbled that the receiver could not even make enough sense of the packet to intelligently ask for a retransmission. If a packet is received with only a minor error, a negative acknowledgement, asking for a retransmission, avoids having to wait for the full timeout period to elapse.

Packet sequencing is only one of a number of possible network protocol–introduced problems that can occur in the packetizing process. The other two most serious problems are the undetected loss of a packet and the duplication of a packet that is successfully transmitted.

Packet Sequencing.   The problem of packet sequencing is a direct result of the hold-and-forward mode of operation, arising from the need to protect each transmission from network-introduced errors. Differential delays along the many paths through the network also introduce the possibility that packets will be received out of proper sequence. In order to protect users, the packets have to be "reassembled" into the same basic message structure they had upon initial transmission into the network. The process of packet reassembly is done at the destination switch—in this case switch 3—using packet sequence information (such as a serial number) which has to be carried through the network along with the user-introduced information.

Losing a Packet.   In Figure 4-3, user A enters the first packet into the network via switch 1. Switch 1 routes packet 1 via switch 2, which receives packet 1 correctly and immediately acknowledges it. However, before packet 1 is transmitted from switch 2 toward switch 3, something goes wrong, and switch 2 fails. Having received an acknowledgement for packet 1, switch 1 is no longer concerned about it.

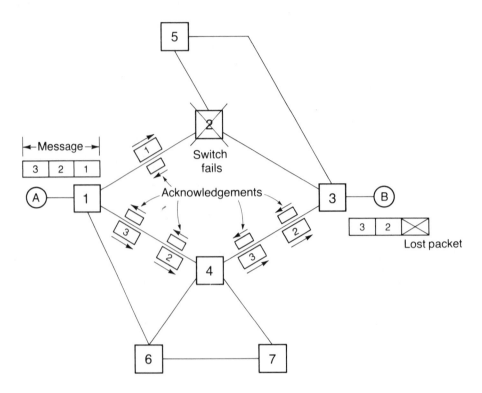

**Figure 4-3.   Basic operation of a packet switched network with a
switch failure: loss of a packet.**

However, upon the failure of switch 2, the network routing plan changes, so that
future packets avoid the failed switch. Thus, packet 2 and packet 3 are routed via
switch 4 and are soon thereafter received properly at the destination switch,
switch 3. But what about packet 1? It was lost when switch 2 failed. Acknowledging
packet 1 meant that switch 1 was no longer responsible for this packet, yet switch 2
failed before it had a chance to relay packet 1 on through the network. For the
destination user B, packet 1 is irretrievably lost, and all he receives are packets 2
and 3. The basic packet switching protocol has thus introduced the possibility of
lost packets.

There are a number of ways the network can protect against this problem.
First, a switch could be restricted from sending an acknowledgement until it has
actually relayed the packet on. Second, ultimate responsibility for the packet could
rest with the originating switch. Third, the sending user could be required to fill
in missing (lost) packets at the request of the destination switch. All these possible
protections, as well as others, have their advantages and disadvantages, which have
to be explored during design and implementation of operational protocols. The

key point is the need to layer multiple protective checks and protocols to deal with the possibility of lost packets.

Duplicate Packets.   In Figure 4-4, the flow of the message from user A to user B is initiated at switch 1 with packet 1 being routed via switch 2. Packet 1 is received correctly by switch 2 and is immediately acknowledged by switch 2. However, just as the acknowledgement for packet 1 leaves switch 2, the line from switch 2 back to switch 1 fails, in the process destroying the acknowledgement for packet 1. Not receiving an acknowledgement for packet 1 within the timeout period causes switch 1 to retransmit packet 1. Detecting that the line between switch 1 and switch 2 has failed naturally causes the network routing plan to send the retransmission of packet 1, together with the transmission of packets 2 and 3, via switch 4.

In the meantime, the first transmission of packet 1 was really received properly by switch 2, even though the acknowledgement of that first copy was never received by switch 1. Having no way to know about the error, switch 2 relays packet 1 to the destination switch, switch 3. Soon thereafter comes the second copy of packet 1,

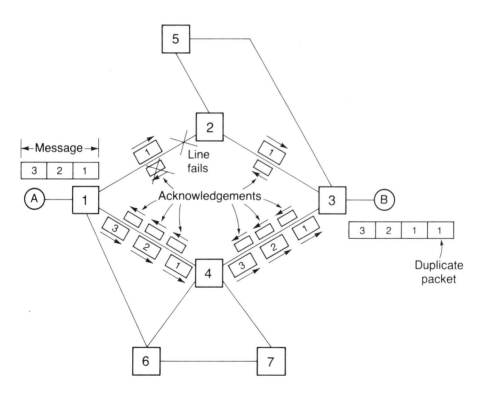

Figure 4-4.   Basic operation of a packet switched network with a line failure: duplication of a packet.

together with packets 2 and 3. What user B is likely to see, therefore, is a duplicate packet 1 together with the rest of the message in its entirety.

Here again, there are numerous ways to avoid the potential problem of duplicate packets. It may not even be a big problem to user B, unless of course the duplicate packet happens to be a second $500 withdrawal from a checking account or a second $200 charge added to a charge account. A simple protocol protection might be to acknowledge successful acknowledgements. But if this is done, it may be necessary to acknowledge the acknowledgements of the acknowledgements, and so on, ad infinitum. Another approach would require the originating switch—node 1 in this case—to put a unique serial number or time stamp on each packet so that the destination switch can check each outgoing packet for possible duplication. The key point is that the protocol is inherently able to cause the creation of duplicate packets, and the network design has to be prepared to deal with and control this possibility.

Network Overhead

We have seen that a typical packet network employing terrestrial switching elements and carrying messages longer than one packet in length may get the message out of sequence, create duplicate packets, or lose packets. Consequently, the internal network operations, as implemented through the switching protocols, have to be clever enough to protect against these conditions. In order to do this, the network has to contain what is called **overhead** information. Overhead is, in effect, capacity that is dedicated to the network's own operation and is not available for the transmission of the user (revenue-producing) information. The overhead information exists in two basic forms: (1) self-contained acknowledgement and control packets flowing among the switches, and (2) overhead that is appended to the user information and flows through the network as part of the information-carrying packets. As overhead information requires a certain percentage of the overall capacity of the network, it has to be designed to keep the useful capacity of the network as great as possible.

Structure.   Figure 4-5 is a general representation of the overhead structure associated with packet switching. User transactions or messages can be of arbitrary length, ranging from just a few bits up to many millions of bits. Because of the way packet networks operate, it is generally not possible, nor is it advantageous, to attempt to transmit the entire message, of length L bits, at one time. The user-to-network interface protocol restricts the user to message or transaction **segments,** which have a maximum length specification of up to M bits. Within the network, the lines and switches exchange packets, each with a maximum length specification of up to N bits. Depending upon the design of the network protocols, the value of M and N may be equal, or M may be permitted to be significantly larger than N. (If M is larger than N, it will generally be an integer multiple of N.) Of course, since the user messages or transactions are of arbitrary length, clearly L is larger

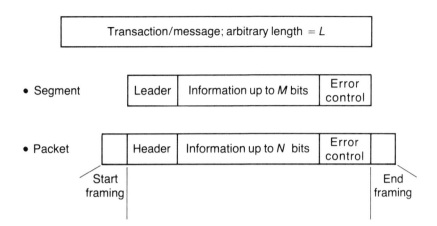

Figure 4-5. Generalized packet switching overhead structure.

than either M or N. Expressed mathematically,

$$L > M \geq N$$

If the network operational protocol requires that M and N be equal, we have what is called a single-packet-per-segment protocol; that is, each segment introduced into the network creates just a single packet. All messages or transactions longer than a single packet have to be divided by the user equipment into single packet–length segments. Where M is permitted to be larger than N, we have what is referred to as a multiple-packet-per-segment protocol. User segments that arrive at the network are of length M; they are then divided by the switches into packets of information length N. The advantage of a multiple-packet-per-segment protocol is that, for user messages of length M bits or less, the packetizing process is a **transparent** operation. That is, if the user always restricts his messages to less than M bits, he does not have to be concerned with the segmentation process, or dividing messages into smaller subunits. It is possible to conceive of protocols where the user-to-network segment length is also arbitrarily large—where, in effect, M and L are permitted to be equal. In such a case, the packet network is actually capable of emulating a full-period point-to-point circuit, with continuous transmission across the network possible.

Types of Protocols.   Examples of each of these types of protocols exist in operational networks. Popular terminology has evolved to categorize these differing modes of operation. Single-packet-per-segment protocols, where each individual packet is handled as a discrete entity, have generally been termed *datagram* networks. Multiple-packet-per-segment protocols, or packet networks that operate on arbitrarily long user inputs without requiring user segmentation, are termed *virtual circuit* packet networks. Virtual circuit implies that, despite the fact that the internal operation of the network is based on packets, the appearance to the end user is indistinguishable from a full-period, end-to-end circuit. The packetized operation must be essentially invisible to the user, with data coming out of the network in exactly the same sequence it went into the network.

True virtual circuit operation is quite difficult to achieve in practice. Because of the need to protect data communications from electronically induced errors during transmission, even full-period, point-to-point circuits generally use a block-by-block mode of transmission, using retransmission to correct errored blocks. Only low- or moderate-speed data terminals, operating in a character asynchronous mode or voice communications, will truly operate in a continuous transmission mode. Nevertheless, the generalized concept of virtual circuit mode of packet network operation is useful, in effect insuring packet sequence and protection from lost and duplicated packets.

User Overhead Compared to Network Overhead

The overhead need of users is largely separable from that of the network. Users interface with the network using segments, whereas the switches interface with each other using packets. The information contained in the segments and the packets may or may not be the same, depending on the protocol in use. However, even with the most simple datagram protocols, it is still likely that there will be significant overhead differences between the user-supplied segments and the network-generated packets.

The user's information segment consists of a segment **leader,** information field, and error-control block (refer back to Figure 4-5). The segment leader contains the destination address to which the segment is to be delivered, together with control information required by the network, such as the segment sequence number, **logical channel number,** designation of the first or last segment of a transaction, and a wide range of protocol information related to the user-to-user control of the circuit. By contrast, the packets that flow among the switches contain framing patterns to designate the beginning and end of the packet, a packet **header,** and an error-control block, in addition to the information supplied by the user. The packet header contains all the same information that the segment leader contains but adds other information needed by the switches to control the movement of the packets through the network. Examples of network information in the packet headers are the source address (in addition to the destination address), packet sequence number (to allow for segment reassembly), control blocks (to prevent looping, in which a packet is routed back and forth among a few switches in an endless circle; loss; or duplication of packets), and various other kinds of information to insure proper operation of the network under a wide range of overload or impaired modes of operation.

A significant amount of information is contained in the segment leaders and packet headers. Typically, anywhere from 64 to 256 bits of total overhead information is required with each packet. In packet networks with the value of N equal to

1000 bits, for instance, 25% of the total data transmitted would be overhead. This 25% of the network capacity is not available to the revenue-producing movement of user information through the network. The potentially large percentage of overhead is, of course, important in comparing the efficiency of packet switching to other communications techniques. From the user's perspective, overall efficiency minimizes the overhead demanded by the network from the user, particularly for very short messages, so as to insure rapid delivery of short, bursty messages through the network.

MATCHING PACKET NETWORKS TO THE BURSTY USER

A large amount of modern telecommunications demand is really very **bursty** in nature; that is, it can be decomposed into a structured sequence of short data bursts separated by comparatively long idle periods. The continuous flow of data through packet networks employing terrestrial switches is achieved by the careful overlaying of a comprehensive set of control protocols to smooth the bursty user information through the continuously available transmission medium.

Switched networks using transmission media of fixed capacity require well-designed protocols in order to achieve efficient operation and high utilization. However, the bursty operation, using media of much higher capacity than the demand of a single user, can be achieved with much simpler protocol operations when network intelligence is built into the user end devices. At the time of the original ARPANET design, data channels operating at 50 kilobits per second were state of the art and very expensive. Digital rates higher than this could be achieved only in localized connections or with very specialized facilities. Rapid improvements in digital communications systems have made a wide range of digital transmission media accessible. Rates as high as many millions of bits per second are now available on such media as satellite channels, optical fibers, coaxial and metallic cable, and digital microwave distribution systems. By combining the higher speed transmission media with processor-controlled devices, the protocol intelligence once constrained to the switching devices can now be readily distributed among the user elements in the network.

Figure 4-6 shows the packet switched network of Figure 4-2, with the discrete switching elements eliminated. The information, rather than flowing on a step-by-step, packet-by-packet, basis from the input user A to the output user B, is communicated by the universal ability of all users to "hear" all packets transmitted. Destination user B achieves the message continuity by the sequential reception of the transmitted packets, using end-to-end error control to insure accurate transmission. Short bursts of information, particularly those contained in a single packet, are rapidly disseminated to the destination users.

The ability to utilize the full capacity of such broad bandwidth interconnection depends on applying any of a number of different broadcast protocols. These broadcast protocols achieve both fully random access to the transmission medium and the flexibility to meet single-packet bursty messages as well as longer messages

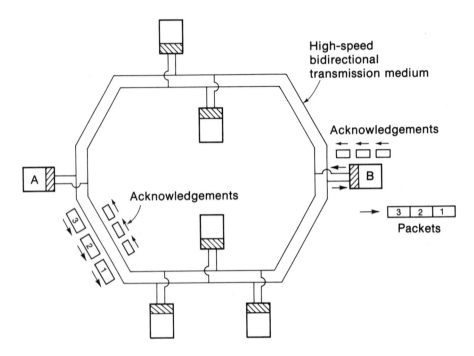

Figure 4-6.   Operation of a packet switched network with discrete switching elements eliminated.

composed of multiple, successive packets. These protocols will be defined and analyzed in subsequent chapters. The key element of their operation is intelligent user end devices, achieving fully distributed network operation. However, as we shall see, some centralized control and management of resources will often result in greater overall efficiency of operation.

COMBINING VOICE AND DATA IN MULTIMODE PACKET NETWORKS

With proper design of the network interface and flow-control protocols, digitized voice can be combined with other digital data services into a common, packet switched network. The integrated voice and data network shown in Figure 4-7 illustrates the point. Voice digitization may take place at the user instrument, at any concentrator in the system (such as an access switch or PBX), or at the input to the packet switch. If the voice processors are placed at a concentration point or at the input to the packet switch, their cost can be shared among many users because the processors can be pooled to handle the number of lines at the switch that are active at any time, rather than the total number of end users in the network.

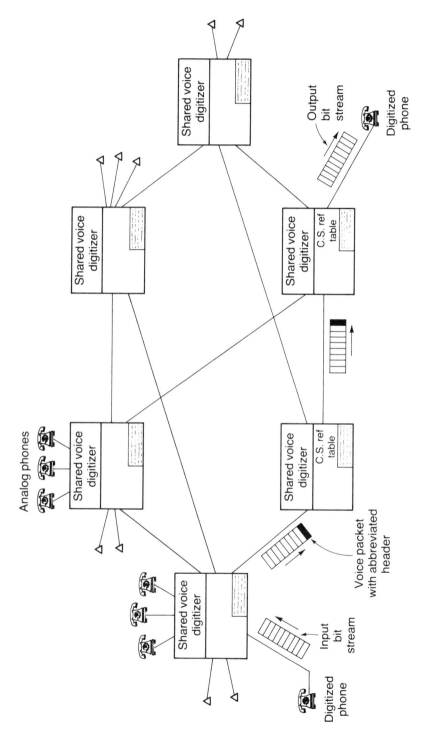

Figure 4-7.   An integrated voice/data packet network.

In order to operate efficiently in the integrated mode, the voice packets have to be uniquely identified (class marked) so they can be handled as expeditiously as possible. Voice packets need not be error checked, since there is no time to retransmit errored voice packets. Packets received with errors will be processed at the output speech synthesizer and will, at worst, result in a short noise burst. Because voice packets contain a relatively small number of bits, it is important to minimize their overhead. This is done by establishing a fixed path through the network during the call set-up, and transmitting all packets associated with that call over the same path. An entry in the circuit switched reference table in the memory of each switch along the end-to-end path reduces the overhead for each packet to the logical circuit reference number. Each switch uses this reference number to determine the proper handling and routing of each packet associated with a particular call.

The overall efficiency of integrating voice and other services into common packet switched networks is related to the ability to create the needed protocols, as well as the cost-effectiveness of the devices used to convert analog voice into digital bit streams. Like so much processor-driven technology, the cost of low-bit-rate voice processors (those at rates of 32,000 bits per second per voice channel, down to possibly as little as 4800 bits per second per channel) will be trivial by comparison to the other aspects of the network within the forseeable future, making the integrated approach even more practical. Moreover, the combination of data entry and voice answer-back functions will also mandate the functional integration of voice and data systems, with packet switching emerging as the most cost-effective approach.

Voice/data integration is also possible with the wideband broadcast techniques illustrated in Figure 4-6. Individual information bursts, whether voice or data, are received and processed on an individually addressed basis. Simplified protocols applied to the voice packets permit expeditious handling of those packets in terms of both error processing and flow maintenance. Other high-redundancy services, such as facsimile, can be similarly handled, with a minimum of error correction to maintain a continuous information flow. All that is required is the ability of the commonly shared resource channel to have (1) a sufficiently high peak data rate to assure the low average delays across the network, and (2) the gross capacity necessary to accommodate the total throughput demanded by the overall user community at peak traffic loading. Terminal processing capability is, of course, also required for fully distributed operation.

SUMMARY

1. Ideally suited to transmit short, individual bursts of user information, the packet switched network can be overlayed with end-to-end flow protocols to allow the transmission of longer, continuous messages or transactions.

2. Protocols also protect the user data from a variety of anomalies, including lost, duplicated, or missequenced pieces of the message.

3. The use of very high-bit-rate shared transmission media can achieve high-capacity operation using broadcast techniques, leading to the ultimate flexibility of fully distributed networks.

4. Voice and other data sources can be integrated in packet networks by simplifying the bursty mode of operation.

SUGGESTED READING

CARR, S., CERF, V., and CROCKER, S. "Host-to-Host Protocol in the ARPA Computer Network." *Proceedings of the Spring Joint Computer Conference*, American Federation of Information Processing Societies (AFIPS), May 1970, pp. 589–597.

Written and published in the earliest days of the operation of the ARPA network, this paper provides an excellent description of the initial host-to-host protocols used in the network. The relationship between the host operating systems and the network nodes is discussed, together with the functions and commands which cross the interface between the two.

GREEN, PAUL E., JR. "An Introduction to Network Architectures and Protocols." *IEEE Transactions on Communications*, vol. COM-28, no. 4 (April 1980), pp. 413–424.

This tutorial paper describes the functions that any network must provide in tying two end users together, and how these lead to the layered approach to network protocol design which has evolved into an international standard. The discussion includes descriptions of several commercial as well as CCITT protocol structures.

POUZIN, LOUIS, and ZIMMERMANN, HUBERT. "A Tutorial on Protocols." *Proceedings of the IEEE*, vol. 66, no. 11 (November 1978), pp. 1346–1370.

This excellent collection of articles, including 85 references on the development and design of networking protocols, emphasizes the user-to-user environment and the relationship between protocols, distributed information systems architecture, and programming languages.

5

Fully Random Access—
The ALOHA Technique

THIS CHAPTER:

will describe the most general approach to an unstructured,
distributed communications system—random broadcasting using
the ALOHA technique.

will describe how ALOHA switching functions and information
exchange take place.

will develop a technical understanding of the channel functioning,
and estimate the capacity, throughput, and delay of systems
based on the ALOHA technique.

With the background of the first four chapters, we can now explore the most general design of a distributed communications network. The universal visibility of satellites over a broad geographic area, combined with the concepts of unstructured multiplexing, narrowcasting, and packet switching, make possible wideranging, distributed, multiuser communications networks that do not require the discrete switching elements found in a terrestrially based network.

THE FUNDAMENTAL ALOHA PROTOCOL OVER A
SATELLITE CHANNEL

The ALOHA technique, or protocol, provides the most fundamental approach to a fully distributed network. The mode of operation of the protocol is suggested by its name, which is used in its literal Hawaiian meaning as a greeting of both arrival and departure. System operation is based upon the users initiating transmission into the common system whenever they have a new message to transmit. In effect, a user says ALOHA—hello—into the communications channel at will, without regard for the current status of the system. There are many refinements to the ALOHA technique, which will be the subject of later chapters; for the most part, they reduce the degree of distribution of the network intelligence.

The Operational Environment for ALOHA Systems

The apparently chaotic operation of the ALOHA technique is premised upon a group of users, each narrowcasting to a central processor (Figure 5-1). The narrowcast takes place over a commonly available satellite channel using a synchronous satellite 22,300 miles above the equator. Because of the long distance the transmissions must travel, it takes about one-fourth of a second for a user's signal to reach the satellite and be returned to earth. Since all users are presumed to have "identical" earth stations, each user can receive the communications from all other users and can also hear his own transmissions. Though the satellite retransmits exactly what it receives, it is more than just a "mirror in the sky." The satellite receives signals transmitted on the **uplink** frequency, amplifies them, and retransmits them on a different **downlink** frequency. All users in the system transmit on the same uplink frequency and listen to the same downlink frequency.

Although the ALOHA technique is not dependent on any particular data rate, it will be easiest to explain the operating principles in terms of a particular set of data rates. Let us assume, then, that the channel is operating at a rate of 50,000 bits per second, and that each user sends his data in packets consisting of 1000 bits or less. A full packet will thus have a duration on the channel of one-fiftieth of a second (20 ms), which is relatively short compared to the quarter-second (250 ms) the packet takes to travel up to the satellite and back.

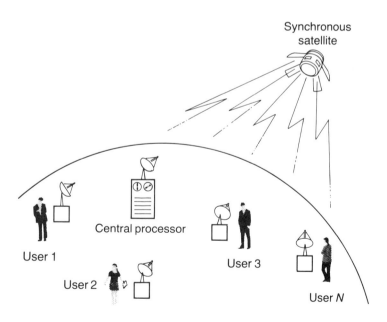

Figure 5-1.   A distributed network with each user narrowcasting via a common satellite channel.

Transmissions and Collisions

Figure 5-2 illustrates the typical occurrences in the ALOHA channel. The figure shows each user's uplink transmissions separately, and, at the bottom, the combined downlink transmission heard from the satellite. First, user 1 transmits a packet lasting 20 ms. Shortly thereafter, user 4 transmits a packet, but before this packet is completed, user 2 begins to transmit a packet. When he begins his transmission, user 2 has no way of knowing that user 4 is currently transmitting because user 2 cannot hear user 4 directly. Moreover, what is heard over the downlink is a quarter-second old and does not accurately represent the current status of the channel. Thus, the packets of users 4 and 2 are destined to overlap; there will be a **collision.** A quarter-second later the collision of these packets is heard in the downlink channel as a pair of individual packets, each with many errors resulting from the mutual interference generated by the collision.

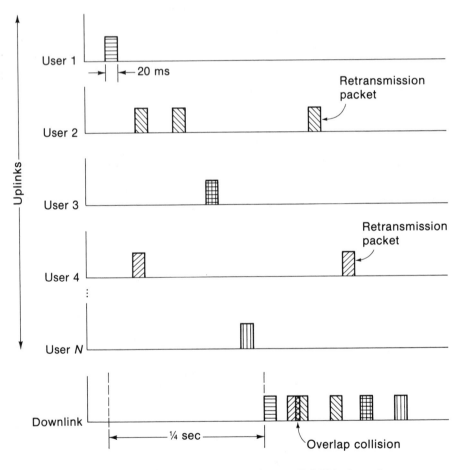

Figure 5-2. Typical occurrences in an ALOHA channel.

It is assumed that, if any part of a packet is damaged, the entire packet has to be retransmitted. It is also assumed that the collision of two packets damages both of them equally, which is a worst case assumption. (Under certain circumstances it is possible for one packet involved in a collision to be received correctly, in effect overpowering the other packet. This situation will be treated in Chapter 6.)

Just as the collided packets are unintelligible at the destination, they are likewise unintelligible to the original sender, who is also listening to the downlink channel. Thus, one quarter-second after initial transmission, user 2 and user 4, both listening to the common downlink, hear that their packets have been involved in a collision, and each transmits a repetition packet to replace the damaged packets. However, if both users act immediately, they are likely to collide again. In order to minimize the probability of a second collision, both users select a random delay time before attempting the retransmission. As we see in Figure 5-2, the procedure is successful, and the retransmission packets of user 2 and user 4 are carried without interference.

ALOHA Channels in Typical Interactive User Situations

The major applications of unstructured resource sharing or multiplex techniques are found in high peak-to-average information situations, best characterized by interactive users, query–response users, or even data-base updates. Common to all these situations is a human user accessing the system by means of a keyboard or sensor device (such as a laser label reader), with reasonably long human thought and reaction times between successive inputs. With human participation in the information input function, the operation of an ALOHA channel portrayed in Figure 5-2 is quite inaccurate from the point of view of the overall time scale. The entire width of the page represents less than 1 second of real time. In reality it is very unlikely—for most users quite impossible—that any one user will generate more than one packet during a 1-second interval. (To do so from a keyboard type of device would require a typing speed in excess of 1500 words per minute.)

In any case, collisions do occur, causing delays and retransmissions. As the number of active users increases, or the frequency with which each user transmits packets increases, the likelihood of collisions increases. As the collisions increase, the channel becomes even busier because each collision generates at least two attempted retransmissions. Therefore, it is important to see if we can predict the behavior of the ALOHA channel and determine how much traffic such a channel is actually capable of delivering.

INFORMATION EXCHANGE USING PACKET NARROWCASTING

Before analyzing the performance of the ALOHA channel, let us focus on some of the unique operational features of the technique that establish the full capabilities of a switched network without the need for switches.

In this discussion the user and his terminal device are considered as the same entity. However, the protocol functions, though described as if they are carried out by the human user, are actually implemented by the user's terminal device. For example, though we say the user must listen for a possible collision one-quarter second after the packet is transmitted, in reality the channel interface equipment associated with the user's terminal would do this and perform any necessary retransmissions. The person at the terminal would not be required to take any actions at all different from the processes and procedures he might follow if he had a dedicated line into the network.

The network functions generally associated with a packet switch are automatically absorbed into the basic operation of the ALOHA channel. In a network employing terrestrial switches, the switches act as the user connection point, performing the packet routing function and the capacity allocation of the lines to which the switch is connected. The switches allocate capacity by buffering packets until each has its turn for transmission over the shared facilities.

Figure 5-1 showed a rather simplistic view of the satellite narrowcast situation, with N users each seeking to transmit packets to a single central computer. Figure 5-3 depicts a situation where users may want to send packets to other users directly, as well as to any one of many computers in the network. Three of

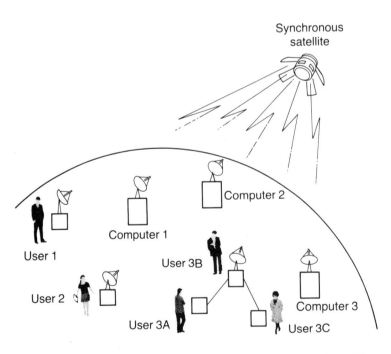

Figure 5-3. A distributed network of users with a shared satellite channel using dedicated and shared earth stations.

the users—user 3A, user 3B, and user 3C—are in close proximity to each other and can share a common ground station equipment set. It is also possible that an entire subnetwork, or local network, can be tied to the satellite channel through a shared interface processor and ground station.

Switching Functions within the Operational Protocols

The switching functions of the ALOHA technique are automatically absorbed into the operational protocols. Network access is achieved by any user who has connection to an authorized satellite earth terminal. A user may be directly connected to such a local earth station, or he may use a short piece of dedicated terrestrial transmission to connect to an earth station. (In a later chapter we will look at the techniques for optimizing the mix of earth stations and terrestrial connectivity to the earth stations.) Routing of traffic is not needed in the way it is for terrestrial packet switched networks since all users transmit directly to each possible destination via the universal broadcast coverage of the satellite. Capacity allocation is automatically achieved by those packets that are successfully transmitted without collisions. All the required capacity, instead of being contained in a large number of individual point-to-point links, is aggregated into the common capacity available through the satellite. If capacity is available when a user transmits, that user gets through successfully on the first try. If the channel is busy and capacity is not available, collisions and retransmission occur until delivery is successful.

The switching functions take place not by the selective transmission and routing of information to the proper destination, but by a process more correctly defined as selective reception of destination traffic. All terminals in the system, by virtue of their ability to "hear" all traffic in the system, are responsible for selectively receiving only that traffic addressed to them. Each packet transmitted, then, requires the inclusion of both the source and the destination address, as well as a repetition counter in order to prevent confusion between original and repeat packets. Each packet must contain all of the needed information to insure its arrival at its ultimate destination. The full range of networking functions are thus achieved using a satellite broadcast technique, without the need for network switches.

Disadvantages of Broadcasting/Narrowcasting

There are, of course, some disadvantages to broadcast network techniques. Since a single satellite can cover only about one-third of the earth's surface, user communities which span very large distances may not be reachable via a single satellite broadcast network. Another problem is privacy. Not only is every user in the network able to hear the traffic of any other user, but anyone, anywhere, within visible range of the satellite can monitor all transmissions of the user community.

Though the problem is somewhat more serious in a broadcast network, all forms of common user or carrier provided communications are vulnerable to interception. As a consequence, users of electronic communications are moving rapidly to utilize security and privacy devices, regardless of the transmission medium.

Other disadvantages concern the maintenance of control and discipline in the network. Operation of the protocol depends on all users adhering to the rules of the protocol and using essentially identical terminal facilities. If one user transmits at higher power than others, that user will "steal" a disproportionate share of the system capacity. When collisions occur, the protocol depends on all users equally selecting a random waiting period for retransmission. In addition, when the channel becomes very busy, additional transmissions result in such a high probability of retransmission that total channel throughput decreases, possibly to a point where almost no useful traffic successfully proceeds through the channel. Such an unstable condition requires at least some external control beyond the basic random broadcast transmission.

Special Advantages

Satellite broadcast and narrowcast techniques provide a number of capabilities that would be much more difficult to implement with conventional switched network techniques. Packets destined for more than one location need to be transmitted only once, with all destinations indicated in the single transmission. Since all users can hear the information at the same time, single transmission of multiple-destination messages reduces capacity utilization. New users can enter the network or existing users can move around without having to rewire facilities or inform any address, routing, or control tables of the changes. Thus, a broadcast-based network can be rapidly reconfigured or expanded and can support mobile users very easily.

When packetized communications follow a satellite broadcast, demand-assignment network configuration, the basic packet switching functions take place inherently in the operation of the channel access protocols. In other words, packet switching is achieved . . . without packet switches!

PERFORMANCE OF THE ALOHA PROTOCOL—
CAPACITY AND DELAY

Analysis

Now that the mode of operation of the ALOHA protocol has been explained, we can apply some relatively simple principles of probability to determine the capacity, throughput, and delay of the channel to user traffic. The notation used to determine the system performance is summarized in Box 5-1.

λ = messages (packets) to be delivered (in packets per second)

λ' = λ + repetition packets

therefore:

$$\lambda' > \lambda$$

K = length of packet (in bits)

R = channel rate (in bits per second)

$$\tau = \text{packet length (in seconds)} = \frac{K}{R}$$

then

$$\lambda \times K = \text{traffic intensity in the channel (in bits per second)}$$

for convenience, define:

$$S = \lambda K = \text{channel throughput}$$
$$G = \lambda' K = \text{channel traffic}$$

or:

$$s = \frac{\lambda K}{R} = \text{normalized channel throughput}$$

$$g = \frac{\lambda' K}{R} = \text{normalized channel traffic}$$

or

$$\boxed{s = \lambda \left(\frac{K}{R} \right) = \lambda \tau}$$

$$\boxed{g = \lambda' \left(\frac{K}{R} \right) = \lambda' \tau}$$

Box 5-1

Let us assume that the aggregate community of users has a total of λ messages, or packets, per second to be delivered at the present time. This is the current packet demand on the channel, measured in packets per second. However, because of the way the ALOHA channel operates, some of the packets that are successfully delivered will have been involved in collisions, resulting in repetition packets. The total traffic actually flowing in the channel is thus defined as λ'—the sum of the packets delivered on the first attempt, the packets delivered as a result of one or more repetitions, and the packets that were damaged by collision. In other words, $\lambda' = \lambda +$ repetition packets. Therefore, though we don't yet know the actual value of λ', it must be larger than λ.

The length of each packet in bits will be denoted by K, and the rate of the channel will be R bits per second. The amount of traffic that the users require to be delivered by the channel is thus $(\lambda \times K)$ bits per second.

For convenience, we will define the following two terms:

$$S = \lambda K = \text{channel } \textit{throughput}$$
$$G = \lambda' K = \text{channel } \textit{traffic}$$

and, since the channel bit rate is R bits per second,

$$s = \frac{\lambda K}{R} = \textit{normalized} \text{ channel } \textit{throughput}$$

$$g = \frac{\lambda' K}{R} = \textit{normalized} \text{ channel } \textit{traffic}$$

By *normalized* throughput and traffic we mean simply that both of these measures are expressed as a fraction of the total capacity of the channel. The values of s and g are thus fractions that can range from 0.0 to 1.0. Furthermore, by expressing the results in a normalized measure, the analysis is completely independent of the actual channel rate that is being considered. However, for purposes of illustration, the basic channel rate will be assumed to be $R = 50,000$ bits per second, with a packet length of $K = 1000$ bits.

A final point concerns packet length, measured not in bits but in time. If the packet length is K bits, and the channel rate is R bits per second, then the time duration of a packet is given as:

$$\tau = \frac{K}{R} \text{ seconds per packet}$$

By substituting these values of τ in the expressions above, we find:

$$s = \lambda \tau$$

and

$$g = \lambda' \tau$$

Determining Capacity

Using this notation, and the definition of the operation of the ALOHA channel, we can determine the capacity of the channel. Figure 5-4 focuses on the transmissions of one individual user of the channel, say user V. Packets are transmitted at random times, each packet lasting τ seconds. However, in order for any particular packet to be received correctly—that is, to avoid collisions—no other user can start to transmit a packet beginning τ seconds ahead of user V or any time during the interval up to the end of user V's transmission. In other words, user V's packet will be successful only if no other user transmits a packet during the interval, which is twice as long as user V's packet.

The **Poisson arrival process (P)** mathematically describes the probability associated with the actions of a large number of statistically unrelated users, on the basis of their average message rates. The probability that a large group of uncorrelated users will generate exactly N new packets during a time interval τ seconds long is given by:

$$P(N) = \frac{(\lambda\tau)^N e^{-\lambda\tau}}{N!}$$

where λ is the average packet arrival rate, e is the base of the natural logarithms, 2.718, and $N! = (N)(N - 1)(N - 2) \ldots (3)(2)(1)$.

This expression is useful because the probability of user V's packet being successful is exactly the same as the probability that exactly zero other packets are transmitted during the time interval of length 2τ seconds. By substituting these values into the Poisson probability expression, the probability of zero transmissions ($N = 0$), which is the probability that user V's packet is successful, is:

$$P_s = P(N = 0) = \frac{(\lambda'2\tau)^0 e^{-2\lambda'\tau}}{0!} = e^{-2\lambda'\tau}$$

The traffic value λ' is used in this expression since, from the point of view of interference with user V's packet, the traffic seen by the channel is the total channel traffic, including all the likely repetition packets. The expression $P_s = e^{-2\lambda'\tau}$ is helpful, but it does not fully describe the utilization of the ALOHA channel.

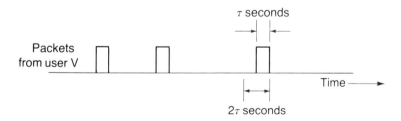

**Figure 5-4. Vulnerable period for packet interference in an
ALOHA channel.**

What the expression does show is the probability that a particular user's packet will be successful when it is transmitted. Another estimate of the probability of success can be derived from the basic definition of probability. In a statistical situation, after a large number of trials, the probability of success can be estimated simply by the ratio of successful outcomes divided by the total number of trials. In the case of the ALOHA channel the average successful traffic is what was defined as the channel throughput, and the total number of trials (attempted packets) was defined as the channel traffic. By the definition of the probability of success, therefore, we have:

$$P_s = \frac{\text{channel throughput}}{\text{channel traffic}} = \frac{\lambda}{\lambda'}$$

Since, on the average, both of these measures of probability of success for the same channel must be equal, these two expressions can be equated, yielding:

$$\frac{\lambda}{\lambda'} = e^{-2\lambda'\tau}$$

or

$$\lambda = \lambda' e^{-2\lambda'\tau}$$

By multiplying both sides of this last expression by the packet length, τ, we change the notation, yielding:

$$\lambda\tau = \lambda'\tau e^{-2\lambda'\tau}$$

and since $\lambda\tau$ and $\lambda'\tau$ define s and g, respectively:

$$\boxed{s = ge^{-2g}}$$

This last expression relates the useful, delivered throughput of the ALOHA channel, s, to the total traffic flowing in the channel, some of which is producing only collisions and no useful delivered traffic. An approximate plot of this relationship is shown in Figure 5-5.

As the channel traffic begins to increase, the useful throughput also begins to increase relatively quickly. However, as the traffic continues to increase, the probability of collisions also increases, resulting in a lower probability of successful transmission. A point is finally reached—at a value of g equal to one-half—where any further increase in traffic creates collisions with such a high probability that the useful throughput is actually reduced. The point of maximum useful throughput, known as the ALOHA channel capacity, occurs at a value of channel traffic of $g = 0.5$. The resulting useful channel throughput is $s = 1/2e = 0.184$. Remember that these were relative, or normalized, measures. Therefore, the maximum useful throughput of the ALOHA channel is only 18.4% of the original channel basic bit rate, and occurs when the channel is filled to 50% of its bit rate with transmitted traffic.

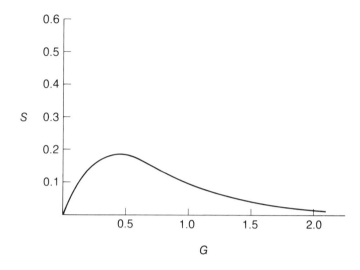

Figure 5-5.   Plot of channel throughput versus channel traffic in an ALOHA channel.

Implications

Although the useful throughput is "only" 18.4% of the basic channel bit rate, that capacity is usable in a very flexible way. Since users have no multiplex equipment and no allocated share of the capacity, the ALOHA channel throughput is composed only of truly useful delivered information. In the case of the earlier example a 50,000-bits per second channel would have a useful throughput of about 9200 bits per second. However, if the true *average* demand of the users was but a few bits per second, as estimated in Table 1-1, the channel could support possibly as many as 5000 users. With the channel parameters shown in Figure 3-3, a channel with a basic rate of 100,000 bits per second would sustain an average rate of 18,000 bits per second. If the average user throughput demand were 2 bits per second, about 9000 users would be accommodated in the single channel.

Such capability is achieved because none of the channel capacity is used unless a user is currently active and ready to transmit. There is no control in the channel other than the repeat protocol for collided packets. Users require no multiplex or timing equipment. The only network overhead is the destination address of each packet since no other network control or routing information is required. The channel capacity is thus used very efficiently, as long as operation is maintained below the channel traffic level of $g = 0.5$.

However, if demand exceeds capacity, such that g begins to exceed a value of 0.5, the additional demand creates more collisions, which in turn create more demand in the form of additional retransmissions, which create more collisions, driving the useful throughput to zero. (As we saw in Figure 5-5, the value of s, the throughput, gets vanishingly small as the value of g, channel traffic, rises above 0.5.) As a result, the users' ability to transmit has to be controlled in order to insure that the operation of the channel always remains in the region below (or immediately around) a value of $g = 0.5$.

The easiest way to achieve control is for a central facility to monitor the channel performance and, as the load becomes heavy, to command the user terminals to increase the time delay before retransmitting a packet that has been involved in a collision. This will reduce the apparent demand and keep operation in the region of high channel throughput. Many other approaches are possible, including the ultimate doomsday approach, which assumes that, as channel performance deteriorates, many users will defer their use to a later time, thus reducing the load to levels that are acceptable to the remainder of the users.

DELAY IN THE ALOHA CHANNEL

The human user of the ALOHA channel sees the collision and retransmission process in terms of the delay between the time a packet is transmitted and the time at which it is ultimately confirmed as delivered to its destination. By using the satellite channel, the user has a minimum delay of approximately 0.25 second to transmit the packet from origin to destination. If an immediate reply or confirmation is expected from the destination, the "round-trip" delay will be a minimum of about 0.5 second, plus the processing time associated with the distant end. If both the user and the destination are using the same ALOHA channel, then the collision/retransmission process will affect both directions of communication.

The Delay Process

Figure 5-6 illustrates the elements of the delay associated with a single packet message through the ALOHA channel. In order to make the discussion general, all times are measured in terms of packet lengths, where τ denotes the length of the packet in seconds. The propagation time up to the satellite and back is given as $N\tau$. For example, since the satellite delay is about 250 ms, for packet lengths of 20 ms N would have a value of 12.5.

The process illustrated in Figure 5-6 proceeds as follows. The packet is initially transmitted, which takes τ seconds. After $N\tau$ seconds of transmission delay the packet is received, possibly with interference. If the packet was received without interference the first time, the added delay of the channel is simply $N\tau$ seconds, the single-hop satellite delay. If a collision occurred, however, the protocol requires that the sender wait for a random period, between zero and K packet times, before retransmitting, in order to minimize the probability of a second collision. On the average, then, users will have to wait $(1 + K)\tau/2$ seconds before attempting retransmission. The retransmitted packet will then proceed over the channel, experiencing a delay of $N\tau$ seconds, and possibly repeating the retransmission process again.

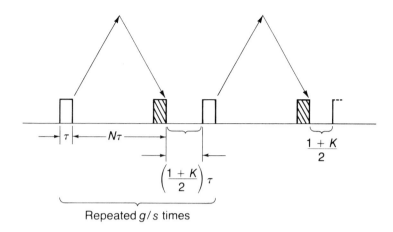

Figure 5-6.   Factors contributing to packet delay in an ALOHA channel.

We can determine the number of times this process has to repeat from the basic definitions of channel traffic and channel throughput. The value of s, representing the channel throughput, is the amount of traffic that is ultimately delivered successfully, regardless of how many times it has to be repeated. The value of g, the channel traffic, is the total traffic in the channel—the successful throughput plus all of the previously unsuccessful collisions. The ratio of g/s, then, represents the average number of times that each packet has to be repeated before being successfully delivered.

For example, at maximum ALOHA channel capacity, the channel of 50,000 bits per second of gross data rate would have a value of $g = 25,000$ bits per second and $s = 9200$ bits per second. The value of g/s, therefore, would be $25,000/9200 = 2.72$; in other words, each bit (and consequently each packet) has to be transmitted an average of 2.72 times. Some packets will make it successfully on the first try, some on the second, and so forth, but on the average it takes g/s transmissions for success. Referring back to Figure 5-6, we can see that the actions covered by the brackets in the figure are repeated an average of g/s times.

We can now write the following expression for the total average delay through the ALOHA channel, including the time to transmit the packet initially:

$$\text{total average delay} = \tau + \left\{ N\tau + \frac{(1 + K)\tau}{2} \right\} \frac{g}{s} - \frac{(1 + K)\tau}{2}$$

The last term in this expression is needed to subtract the last random wait before retransmission for the packet that is finally successful.

We can express the total average delay in a somewhat more convenient form:

$$\text{total average delay} = \frac{\tau}{2} \left[1 + e^{2g}(1 + 2N) + K(e^{2g} - 1) \right]$$

$$\text{TAD} = \text{total average delay} = \tau + \left\{ N\tau + \frac{(1 + K)\tau}{2} \right\} \frac{g}{s} - \left(\frac{1 + K}{2} \right) \tau$$

from the ALOHA capacity equation:

$$s = ge^{-2g}$$

therefore

$$\frac{g}{s} = \frac{1}{e^{-2g}} = e^{2g}$$

$$\text{TAD} = \tau + \left\{ N\tau + \frac{(1 + K)\tau}{2} \right\} e^{2g} - \left(\frac{1 + K}{2} \right) \tau$$

$$= \tau + N\tau e^{2g} + \frac{K\tau}{2} e^{2g} + \frac{\tau}{2} e^{2g} - \frac{K\tau}{2} - \frac{\tau}{2}$$

$$= \tau \left[\frac{1}{2} + \frac{e^{2g}}{2} + \frac{K}{2}(e^{2g} - 1) + Ne^{2g} \right]$$

$$\boxed{\text{total average delay} = \frac{\tau}{2} \left[1 + e^{2g}(1 + 2N) + K(e^{2g} - 1) \right]}$$

where

τ = packet length (in seconds)

g = channel traffic factor

N = satellite propagation delay (in packet lengths)

K = retransmission protocol delay (in packet lengths)

e = base of natural logarithms = 2.718

Box 5-2

where the first parenthetical term relates to the satellite delay, and the second parenthetical term relates to the retransmission protocol waiting delay. Box 5-2 presents the mathematical derivation of this expression. Graphical presentation and some illustrative examples may make it easier to understand these concepts.

Examples of Delay Computation

The plot in Figure 5-7 is based on the parameters used in earlier ALOHA channel examples. Assume a basic channel rate of 50,000 bits per second and a

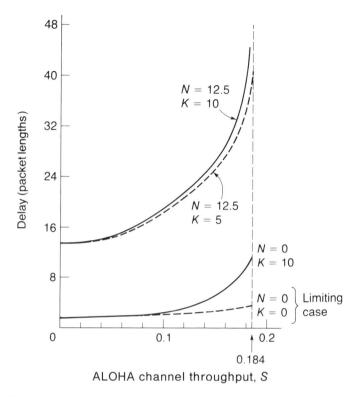

N = Round-trip satellite delay,
    in packet lengths

K = Protocol retransmission
    delay (maximum waiting
    time before retransmitting
    a collided packet), in
    packet lengths

**Figure 5-7.   Plot of delay versus channel throughput for the basic
ALOHA channel.**

packet length of 1000 bits. Thus, the value of τ is 20 ms, and the satellite round-trip delay is $N = 12.5$. Figure 5-7 plots the user-observed average delay (in packet lengths) as a function of the channel throughput. The curves are plotted only up to the value of maximum channel throughput, corresponding to a value of $s = 0.184$.

When the throughput is low, the traffic is very light in the channel and there are very few collisions, resulting in low values of delay. As traffic throughput increases, the channel is more heavily loaded, resulting in more collisions, more retransmissions, and more average delay. The upper curve in the figure is based upon values of 12.5 packet lengths satellite propagation delay and 10 packet lengths maximum protocol retransmission waiting. The middle curve shortens the retransmission waiting time to a maximum of 5 packet lengths. The lower curve treats the theoretical case of $N = 0$, or zero propagation delay, which would be the case for a terrestrial repeater system (discussed in more detail in Chapter 7).

Box 5-3 shows the specific calculation for average delay of a 50,000-bits per second ALOHA channel, operating at capacity ($s = 0.184$, $g = 0.5$), with 1000-bit packets ($\tau = 20$ ms). The delay of the ALOHA channel operating at maximum capacity is less than 0.9 second for a 1000-bit packet in the 50,000-bits per second channel. At this level of performance the actual throughput of such a channel is about 9200 bits per second. If the average user demand over time is, for example, a full packet every 2 minutes, the channel could support more than 1100 users, with each user seeing an average delay of about 0.9 second. If the demand or average usage were smaller, then the delay would be less, and the number of users who could be accommodated would be larger. At an overall average demand of, for example, 2 bits per second per user, the channel would be able to accommodate nearly 5000 users.

$$\mathbf{TAD} = \text{total average delay} = \frac{\tau}{2}\left[1 + e^{2g}(1 + 2N) + K(e^{2g} - 1)\right]$$

where:

$$\tau = 20 \text{ ms} = \frac{1000 \text{ bits}}{50,000 \text{ b/s}} = \text{packet length}$$

$$g = 0.5 \text{ at maximum capacity}$$

$$N = 12.5 \text{ packet lengths}$$

$$K = 10 \text{ packet lengths}$$

$$\mathbf{TAD} = \frac{20}{2}\left[1 + e^{2(0.5)}(1 + 2(12.5)) + 10(e^{2(0.5)} - 1)\right]$$

$$= 10[1 + 2.718(1 + 25) + 10(2.718 - 1)]$$

$$= 10[1 + 70.67 + 17.18] = 10[88.85]$$

$$= 888 \text{ ms} \approx 0.9 \text{ second}$$

Box 5-3

SUMMARY

1. In a random-access, assigned-demand, broadcast channel, where any user is permitted to transmit whenever he cares to, demand conflicts for the common capacity are resolved by each user repeating his transmission until it is successful.

2. By using a satellite channel in broadcast operation, each user can "hear" the transmissions of every other user and can therefore hear if any other transmission will cause interference with his own transmission. However, since there is a quarter-second of delay inherent in the satellite channel, a quiet channel now does not necessarily mean that it really is free of other traffic.

3. The nature of the satellite broadcast channel, with universal access from all user locations, achieves most of the features and services of a packet switched network without packet switches. Destination and source address information permits the user packets to be delivered to the destination with an acknowledgement returned to the source.

4. Though the overall capacity of the ALOHA channel is only about 18% of the basic channel rate, the simple allocation protocol, together with the fact that only *active* users consume any of the channel capacity, results in the channel being able to support a much larger number of users than a more structured allocation of the same basic channel rate.

5. Finally, the overall average delay through the channel is quite acceptable (less than 1 second for typical cases) as long as channel operation is maintained in the region below full capacity.

SUGGESTED READING

ABRAMSON, NORMAN. "The Throughput of Packet Broadcasting Channels." *IEEE Transactions on Communications*, vol. COM-25, no. 1 (January 1977), pp. 117–128.

Bringing together the results of several researchers, this paper presents a unified development of the various aspects of packet broadcasting. The presentation treats not only the basic ALOHA channel but also several variations in the ALOHA protocol and the application of packet broadcasting to terrestrial networks. The paper concludes with a summary of some practical applications of packet broadcasting.

KLEINROCK, LEONARD. *Queueing Systems, Volume II: Computer Applications.* New York: John Wiley, 1976, pp. 360–407.

Kleinrock presents a different approach to the analysis of packet broadcasting channels, placing particular emphasis on some of the techniques that can increase the apparent channel capacity. The discussion of ground radio packet switching techniques provides an excellent overview of the carrier sense multiple access techniques, which achieve high channel capacity and utilization.

6

Capacity Improvement by Slotted
ALOHA and Reservations

THIS CHAPTER:

will describe methods by which the basic capacity of the ALOHA
channel can be increased by applying somewhat more discipline
to the users' operation.

will analyze the slotted ALOHA technique, which permits users to
transmit only at discrete time intervals.

will determine the effects on ALOHA channel capacity when
some packets survive a collision.

will introduce a number of other algorithms that "preassign"
capacity and can reach nearly 100% utilization of total channel
capacity.

The operation of the satellite packet broadcast channel using the basic
ALOHA protocol permits users to transmit packets whenever they care to, in
a purely random fashion. As a result, there is a relatively high probability that at
least some portion of any given packet will be interfered with by some other
user's packet. In analyzing the channel operation, we found that the frequency of
packet collisions results in an overall channel capacity of about 18% of the basic
channel rate. Moreover, if the users attempt to use the channel beyond this rate,
the probability of collision becomes so great that the throughput on the channel
quickly tends toward zero.

INCREASING THE CAPACITY OF THE
PACKET BROADCAST CHANNEL

The apparent lack of efficiency of the basic ALOHA channel has stimulated
a great deal of research into channel operations that will allow for greater capacity
while retaining the inherent simplicity and ease of the random broadcast protocol.

Many possible techniques have been suggested, but each involves increased complexity of both the access protocol and the user terminal devices and channel control equipment. Nevertheless, the very low incremental cost of additional complexity in the form of microprocessor devices within user terminals makes many of these approaches very attractive.

The Slotted ALOHA Channel

The **slotted ALOHA channel** protocol decreases the probability of interference between packets by requiring that users transmit only at the beginning of discrete time intervals. This kind of channel protocol means that two users can interfere with each other only if they transmit at exactly the same time. If only one user transmits at the beginning of a packet interval, his packet is "guaranteed" no interference since no other user is permitted to transmit until the beginning of the next packet interval. This technique is capable of effectively doubling the channel capacity at only a small increase in average delay.

Capture Effect in the ALOHA Channel

Improvement in the capacity of an ALOHA channel occurs if each user transmits at a slightly different power level. If two packets with different signal levels collide at the satellite, the stronger of the two signals is likely to **capture** the receiver and be transmitted by the satellite without error. Thus, when a collision occurs, only one of the two colliding packets will need to be retransmitted. Such a protocol could be implemented in two ways. First, power levels could be adjusted upward or downward on a random basis. Or the power levels could be determined on a priority basis, with more important or higher priority users being given higher powered terminals, thus establishing a multiclass priority packet broadcast system. In either case, a slotted ALOHA channel with capture will result in a channel capacity more than three times as large as the 18% of the basic ALOHA technique.

High Efficiency through Capacity Reservation

There are many ways in which much of the randomness of the packet broadcast channel can be eliminated by requiring users to reserve the capacity they are intending to use. So-called **reservation techniques** are particularly useful in increasing the capacity of the packet broadcast channel when some of the users consistently have more information to transmit than will fit into a single packet—for example, when many low-capacity or low-utilization terminals are trying to transmit to a central point while the central processor or computer is sending a relatively large amount of data back to the individual users. Use of slot reservations in the broadcast channel permit these two kinds of traffic to be mixed together, without mutual interference, in such a way as to achieve very high channel capacity. A number of different reservation techniques will be discussed later in this chapter.

THE SLOTTED ALOHA CHANNEL—CAPACITY AND DELAY

The basic, or unslotted, ALOHA channel that was analyzed in the last chapter is a marvel of simple protocol operation (neglecting, of course, the technological achievement of placing the satellite in orbit to begin with). However, inherent in the technique is a period of time, equal to twice the length of a packet, over which each packet is vulnerable to interference. By establishing a slotted channel—that is, a channel with discrete time slots in which users transmit their packets—the vulnerability to collision or interference is considerably reduced.

Establishing Slots on the ALOHA Channel

Figure 6-1 depicts the slotted ALOHA system. Imagine a master timing clock at the satellite or at a master earth station used to synchronize all the user terminals. The clock ticks at intervals exactly corresponding to the length of a packet (for example, every 20 milliseconds for a 50,000-bits per second channel

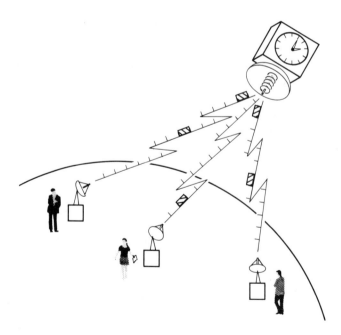

Figure 6-1.   ALOHA channel configuration using a slotted channel.

using 1000-bit packets), and users are allowed to transmit only when the clock ticks. Each user terminal can hear the clock ticking by listening to the channel, and synchronizes his transmitter to the clock. The transmission protocol requires that, when a user has a packet to transmit, he must wait until the beginning of the next time interval—that is, until the next clock tick—before he can actually transmit the packet. If users follow this discipline, packets overlap either completely or not at all. Packets are not destroyed or damaged by having just a few bits at the beginning or end overlapped by another user's earlier or later packet, as can happen with the unslotted ALOHA protocol. In Figure 6-1 the two packets closest to the satellite are about to collide, but the remaining packets in the system are safe from collision.

Disadvantages

The slotted ALOHA channel has two disadvantages. One is the potential complexity of establishing the synchronized time reference for all users in the system. Not only must the terminals be able to provide for the time reference, they must also allow for the slight variation in actual distance between each terminal and the satellite, depending on the exact location of the terminal within the coverage area of the satellite over the earth's surface. Since the time reference (the exact time of the ticks) has to be precise, each terminal must be able to determine range or distance and adjust its transmit time on the basis of its true range compared to a network standard.

Another disadvantage is the fact that the packet length, and the resultant time between clock ticks, represents the maximum amount of data that a single user can transmit at a given time. However, users often have much less information to transmit than the amount permitted in a single packet. For example, a 1000-bit packet, which would correspond to 20-millisecond time slots on a 50,000-bits per second channel, represents about 125 characters (or approximately two standard typewritten lines). Typical user data terminal operations—such as inputting a request or inquiry into a central computer system—often transmit only a small percentage of this amount of data at any given time. In such a case the time between the end of a user's transmission and the beginning of the next time slot is wasted.

Analysis of the Slotted ALOHA Channel

The analysis of the basic or unslotted ALOHA channel developed in Chapter 5 was really a worst case analysis, assuming that every transmitted packet was of maximum allowable length and thus presented the maximum likelihood of interference. The analysis for the slotted ALOHA case is really a best case analysis, assuming that every packet is filled to maximum length and thus does not waste any capacity between the end of one packet and the beginning of the next allowable time slot.

Box 6-1 shows the analytical situation for the slotted ALOHA channel. We begin with the following definition:

s_i = the probability that user i successfully transmits a packet

g_i = the probability that user i transmits any packet (that is, a successful packet or unsuccessful packet)

For user i to transmit a successful packet, since this is a slotted case, he must transmit his packet while no other user transmits one. Mathematically, this can be stated:

$$s_i = g_i(1 - g_1)(1 - g_2)(1 - g_3) \cdots (1 - g_n)/(1 - g_i)$$

All terms are included in this equation for all n possible users of the channel. The expression is divided by $(1 - g_i)$ because we want to exclude the term involving user i since it is this user's probability of success that we are attempting to compute. Mathematically, the equation says simply that the probability of user i's packet being successful is the product of the probability that user i transmits any packet multiplied by the probability that every other user (such as user j) in the system does not transmit any packet. Using the symbol $\prod$ to indicate the product of terms, this expression can be written:

$$s_i = g_i \prod_{\substack{j=1 \\ j \neq i}}^{j=n} (1 - g_j)$$

It is difficult to proceed beyond this point without making an assumption that was inherent in the analysis of the unslotted ALOHA channel: All users of the channel are assumed to be statistically equal; that is, they share the capacity equally, and each user has an equally likely probability of transmitting at any given instant. Under this condition the following expressions can be written:

$$s_i = S/n \quad \text{and} \quad g_i = G/n$$

where S and G are the channel throughput and channel traffic, respectively, with the same meaning they had in the unslotted analysis.

These expressions say simply that a given user's probability of a successful packet, s_i, is one-n^{th} of the total normalized throughput of the channel. Similarly, the probability of user i making a transmission is just one-n^{th} of the total traffic transmitted on the channel.

By substituting these expressions in the equation for the probability of packet success, we find that:

$$S/n = G/n \prod_{\substack{j=1 \\ j \neq i}}^{j=n} (1 - G/n) = G/n(1 - G/n)^{n-1}$$

or

$$S = G(1 - G/n)^{n-1}$$

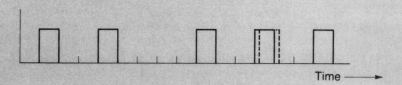

$$n = \text{total number of users in the system}$$
$$s_i = \text{probability of a successful packet by the } i^{\text{th}} \text{ user}$$
$$g_i = \text{probability that the } i^{\text{th}} \text{ user transmits a packet}$$

thus

$$(1 - g_j) = \text{probability that the } j^{\text{th}} \text{ user does not transmit a packet}$$

$$s_i = g_i \prod_{\substack{j=1 \\ j \neq i}}^{j=n} (1 - g_j)$$

for equal users

$$s_i = \frac{S}{n} \quad \text{and} \quad g_i = \frac{G}{n}$$

$$\frac{S}{n} = \frac{G}{n} \prod_{\substack{j=1 \\ j \neq i}}^{j=n} \left(1 - \frac{G}{n}\right) = \frac{G}{n}\left(1 - \frac{G}{n}\right)^{n-1}$$

$$S = G\left(1 - \frac{G}{n}\right)^{n-1}$$

if n is large,

$$\left(1 - \frac{G}{n}\right)^{n-1} \approx e^{-G}$$

thus

$$S = Ge^{-G}$$

Box 6-1

This last expression results because we have the product of $n - 1$ identical terms, each of the form $(1 - G/n)$.

If the number of users is large, we can employ a limit property for the expression $(1 - x/n)^{n-1}$, which is approximately e^{-x} as n tends toward infinity. As a result, the equation for S, for many users, becomes:

$$S = Ge^{-G}$$

Recall that the result for the unslotted case is:

$$S = Ge^{-2G}$$

The results of both these expressions are shown in Figure 6-2, where the relationship of channel throughput to channel traffic is plotted for both the slotted and unslotted ALOHA cases. Notice that, for any given level of channel traffic, the slotted case is better than the unslotted case by a small amount, up to the point where the unslotted channel reaches maximum capacity ($G = \frac{1}{2}$). At this point the unslotted throughput begins to decline, while the throughput of the slotted channel continues to increase. The curve for the slotted channel ultimately reaches a value of $1/e$, or about 0.37, representing a normalized throughput of about 37% of the basic channel rate. This throughput is exactly twice the achievable throughput of the unslotted ALOHA channel.

We must realize that the impression that the channel is achieving twice the throughput is based on the presumption that each packet is full. If the packets are, on the average, half full, then the actual throughput is back to the same value as that for the unslotted channel since no other user can transmit during the empty interval after a partially filled packet ends and the next slot time begins.

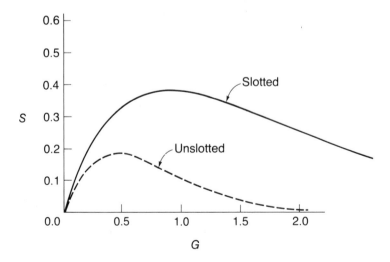

Figure 6-2.   Plot of channel throughput versus channel traffic for slotted and unslotted ALOHA channels.

Deriving Delay

The delay associated with the slotted ALOHA channel can be derived in a fashion analogous to the unslotted delay as was determined in Chapter 5. The only difference is that, on the average, each time a user is ready to make a transmission, he has to wait one-half of a packet time until the beginning of the next slot interval before he can actually transmit. At peak capacity $G = 1$ and $S = 0.37$, which means that each successful packet is transmitted an average of 2.7 times, and each retransmission results in an average of one-half packet time delay, compared to the actual time that the user is ready to transmit. Thus, the maximum excess delay for the slotted channel turns out to be about 1.4 packet times longer than for the unslotted channel operating at maximum capacity. For the 20-millisecond packet time this excess delay—less than 30 milliseconds—is negligible compared to the approximately 900-millisecond delay encountered by the average packet when the channel is operating at maximum capacity.

Application of the Slotted ALOHA Channel

The slotted ALOHA packet broadcasting channel can achieve a twofold increase in channel throughput capacity at a slight increase in average delay, assuming that all packets are filled to their maximum allowable length. The slotted technique requires some additional complexity in user equipment, but with some additional processing within the user terminals, the incremental cost should be low. The slotted ALOHA approach is ideally suited for a highly homogeneous user community where all transmissions of all users are identical in length, although they occur at random times. The packet length could be set to the exact length of each user's transmission, and no capacity would be wasted with partially filled packets. For example, a network of credit card validation terminals or electronic funds transfer terminals, where every transmission represents the same amount of data, might meet this description.

SLOTTED ALOHA CHANNEL WITH CAPTURE

The analysis of the ALOHA packet broadcast channel assumed that, when any part of two or more packets overlap, all packets involved in the collision must be retransmitted. In reality, there is at least some probability that one of the packets involved in a collision will be sufficiently strong to capture the receiver and be received accurately. If this were the case, not every packet involved in a collision would have to be retransmitted, which would reduce the apparent interference and increase the channel throughput at any level of traffic. In such a case it would be possible to create a priority-based system, where users with higher need to communicate can be assigned greater transmitter power, giving them a substantially higher probability of being received correctly even in the presence of interfering packets.

We can analyze this situation initially by assuming sufficient random fluctuation in received signal levels that, for any pair of users, each one has a one-half probability of capturing the channel and being received correctly. We will also assume that, if three users collide, none of the three has sufficient strength to dominate the other two, and thus all three (or more) packets are lost.

Accounting for channel capture requires only a slight modification of the analysis for the slotted ALOHA case. On a slotted channel, assuming that one user captures the channel if exactly two users collide, the success probability for user i can be expressed as:

$$s_i = g_i \prod_{\substack{j=1 \\ j \neq i}}^{j=n} (1 - g_j) + \frac{1}{2} \sum_{\substack{m=1 \\ m \neq i}}^{m=n} g_i g_m \prod_{\substack{k=1 \\ k \neq i \\ k \neq m}}^{k=n} (1 - g_k)$$

$$s_i = g_i \prod_{\substack{j=1 \\ j \neq i}}^{j=n} (1 - g_j) + \frac{1}{2} \sum_{\substack{m=1 \\ m \neq i}}^{m=n} g_i g_m \prod_{\substack{k=1 \\ k \neq i \\ k \neq m}}^{k=n} (1 - g_k)$$

$$s_i = \frac{S}{n} \qquad g_i = \frac{G}{n}$$

$$\frac{S}{n} = \frac{G}{n} \left(1 - \frac{G}{n}\right)^{n-1} + \left(\frac{n-1}{2}\right) \frac{G^2}{n^2} \left(1 - \frac{G}{n}\right)^{n-2}$$

$$S = G \left(1 - \frac{G}{n}\right)^{n-1} + \frac{n-1}{2} \frac{G^2}{n} \left(1 - \frac{G}{n}\right)^{n-2}$$

$$S = G \left(1 - \frac{G}{n}\right)^{n-1} \left[1 + \frac{n-1}{2} \frac{G}{n} \frac{1}{\left(1 - \frac{G}{n}\right)}\right]$$

$$S = G \left(1 - \frac{G}{n}\right)^{n-1} \left[1 + \frac{\left(1 - \frac{1}{n}\right) G}{2 \left(1 - \frac{G}{n}\right)}\right]$$

As n gets very large, $n \to \infty$

$$S = G e^{-G} \left[1 + \frac{G}{2}\right]$$

Box 6-2

The first part of this expression is exactly the same as for the slotted ALOHA case. The second part, which sums up the results of many products, takes the probability that user i will transmit at the same time as one, and only one, other user, and considers each other user in turn. Multiplying this probability by $\frac{1}{2}$ allows for the fact that user i will be successful in the collision half the time, and the other user will be successful half the time.

As in the previous section, for n identical users, the following substitutions can be made:

$$s_i = S/n \quad \text{and} \quad g_i = G/n$$

resulting in

$$S/n = G/n(1 - G/n)^{n-1} + \tfrac{1}{2}(n - 1)G^2/n^2(1 - G/n)^{n-2}$$

or

$$S = G(1 - G/n)^{n-1} + \tfrac{1}{2}(n - 1)G^2/n(1 - G/n)^{n-2}$$

By grouping and rearranging the terms in the expression, as shown in Box 6-2, we can arrive at an expression that can be approximated, for a relatively large value of n (that is, a system with a large number of equal users), by:

$$S = Ge^{-G}(1 + G/2)$$

This expression is plotted in Figure 6-3, together with the capacity curves for the unslotted and slotted ALOHA cases. Not unexpectedly, the capture effect

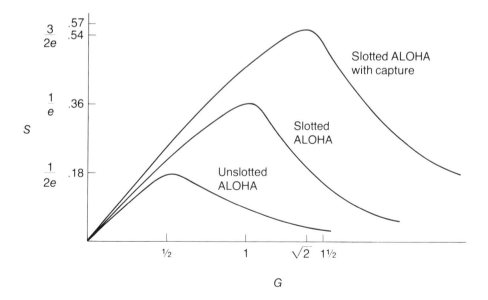

Figure 6-3.   Plot of throughput versus channel traffic for slotted, unslotted, and capture ALOHA channels.

has permitted a roughly 50% increase in channel capacity to an overall capacity of about 57% of the gross channel bit rate. Furthermore, the value of capture probability can be easily changed by simply changing the denominator in the last term of the capacity expression. For instance, if in actual operation it is found that, when two packets collide, one-third of the time user i gets through, one-third of the time user j gets through, and one-third of the time neither of the two gets through, the collision success probability can be adjusted to one-third. The capacity expression then becomes:

$$S = Ge^{-G}(1 + G/3)$$

which naturally results in a throughput reduction compared to the case where each of the users gets through with a probability of one-half.

RESERVING CAPACITY ON PACKET BROADCAST CHANNELS

In analyzing the satellite broadcast channel in its various modes of operation, we made many ideal assumptions in order to get results that could be readily expressed mathematically or graphically. We assumed that the user community was very large (n approaching infinity) and that the packet origination obeyed Poisson statistics. Furthermore, we assumed that the users were functionally identical; that is, they behaved in the same statistical fashion and shared the useful channel throughput equally. Though it is not particularly useful at this point to resolve the mathematics of these assumptions, it will be enlightening to discuss some of the real-world situations that affect the operation of actual systems based on these principles. One response to real-world conditions is the use of capacity reservation techniques.

Unbalanced Users in a Slotted
ALOHA Channel

An unbalanced user of a packet broadcast channel is one with much more traffic to transmit than the average user. Imagine a slotted ALOHA channel with no traffic in it and only one user in the system currently active. The active user has a great deal of traffic and begins to transmit in every time slot of the channel. Since he is the only active user, he never interferes with his own traffic, and therefore achieves a throughput of 100% of the basic channel rate, even though the theoretical maximum throughput of the slotted ALOHA channel is 37%. Clearly, the assumption of many small users is not correct when only one large user is operating in the channel.

Now let us assume there is one large user in a slotted ALOHA channel who has an average demand of 80% of the basic channel rate. Even if this user transmits so often as to consume 80% of the channel capacity, in theory at least part of the remaining 20% is available to other users. We will assume that there are also many small users, whose aggregate demand is 5% of the basic channel rate. The group of small users operate in the channel as if it were the same slotted ALOHA channel that was described earlier in the chapter. The channel now sees a mixture of traffic demands, totaling 85% of the basic channel rate. Since 80% of that demand comes from a single user, who never interferes with himself, there is a good chance that the channel will be capable of satisfying the total demand—again far exceeding the theoretical capacity of 37% of the basic channel bit rate for a large number of identical, small users. In fact, we can view this channel almost as if it were two separate, parallel channels. One is a dedicated channel of 80% of the basic channel rate, operating at 100% efficiency. The other channel comprises the remaining 20% of the basic capacity, carrying the remaining 5% of total demand. The efficiency on the small channel is only 25% (5%/20%), well within the slotted ALOHA prediction for many small users. This is only a rough approximation of the mixture of large and small demand users in a single slotted ALOHA channel, but it provides a useful viewpoint.

Mixing Unbalanced and Small Users

Without trying to compute exact boundaries, we can approximate channel performance curves for a slotted ALOHA channel with mixed large and small users, as shown in Figure 6-4. This curve shows the total channel throughput, S,

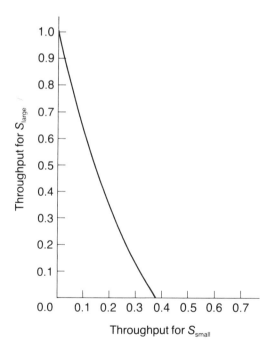

Figure 6-4.   Throughput for a few large users versus many small users of a slotted ALOHA channel.

as the mixture of throughput from a single large user, S_{large}, and many small users, S_{small}. When all the traffic is from a single large user, a throughput of 1.0, or 100% of the basic channel rate, can be achieved. When the throughput is entirely due to small users, a throughput of 0.37, or 37% of the basic channel rate, is achieved.

But what about the more general situation in which the channel is used by many users, some of whom are large, some of whom are small, and some of whom fall in between? Worse yet, what if the users of the channel normally have very little to transmit, so that their traffic generally fits into a single packet, but every so often they have much more information to transmit? With slight increases in the complexity of the channel protocols, these cases can be accommodated very efficiently by use of reservation techniques.

Reserving Capacity by Implication

The first reservation technique of interest is based on the assumption that, once a user starts to transmit, there is a good chance that he will have more than one packet's worth of data. To satisfy this assumption, a master frame structure is superimposed on the slotted ALOHA channel, as illustrated in Figure 6-5. The length of the master frame is R time slots (packet times), such that the total length of the frame exceeds the round-trip delay to the satellite. The beginning of each frame is indicated by a synchronizing preamble transmitted by the satellite or a master earth station. Each packet slot in the frame is numbered, one through R, and each terminal is required to keep track of which of the packet slots were used

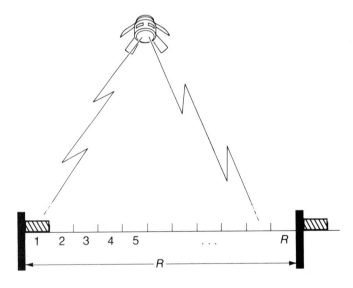

Figure 6-5.   Frame structure superimposed on a slotted ALOHA channel: Total frame length, R, exceeds satellite round-trip delay.

during the last frame. If a user desires to start transmitting, he must select a packet slot that was unused during the last frame.

The only way a collision can occur is if two users pick exactly the same vacant slot during the same quarter-second frame period and begin transmitting at the same time. Once a user successfully transmits a packet in a vacant time slot, he may continue to use that time slot in every successive frame without further interference. When he has completed his transmission, he leaves the slot empty in the following frame. Hearing the slot empty, the remaining users can now select that slot to fill their subsequent demands in later frames.

As a specific example, let us return to the basic channel rate of 50,000 bits per second, with 1000-bit packets, each lasting 20 milliseconds. If the value of R is set to 12, there will be 12 packet intervals in each frame, times 20 milliseconds each, totaling 240 milliseconds, which is essentially the round-trip delay to the satellite. If the channel is lightly loaded, most of the slots will be empty, and the channel will operate like the slotted ALOHA channel we studied earlier, with a maximum capacity of 37% of 50,000 bits per second, or about 18,500 bits per second. But if one large user becomes active, with a substantial amount of traffic, he will transmit full packets in every frame, each time using the same slot. Since there are about four frames per second, he will be achieving a throughput of about 4000 bits per second, without any collisions, while the remaining 11 packet slots are being used in the slotted ALOHA fashion.

As the channel becomes busier with high-volume users, we will eventually reach the condition where there are 12 high-volume users, each in one of the time slots and each achieving an average throughput of 4000 bits per second. The remaining users are forced to wait until one of the active users stops transmitting. The channel in this condition has become in effect a time-division, multiple access channel. It is operating at essentially 100% capacity, but it is now serving only 12 users, not the larger community that was intended. Hopefully, such a condition would not occur too often.

A "Fairer" Approach to Reserving Slots

A variation of this reservation technique avoids the problem of the channel being locked up with a few large users, out of reach of the small users. Instead of R packet slots in the frame, there are N slots, where N is the total number of users in the network. Of course, this makes the overall length of the frame much longer than the round-trip delay to the satellite.

Each user is assigned his own slot in the frame, and he must initiate his transmissions in this slot. However, if a user hears that some other user's slot was empty on the previous frame, he is permitted to assume control of the slot so he can increase his total available capacity. Once a given user assumes the use of another's slot, no other user is permitted to try to use that slot. However, if the rightful owner of the slot has information to transmit, he can regain control of his own slot simply by initiating transmission in it. The first time he does this, it will cause a collision with the borrower's traffic, which, one round-trip delay later, is heard by the borrower of the slot. The borrower is then required to cease transmission, and, after one round-trip delay, the slot is returned to its rightful owner. The only capacity lost to collisions using this protocol comes from the intentional collisions used to signal the return of a packet slot to its rightful owner.

Slot Reservation by Advance Booking

Probably the most efficient approach to dynamic packet reservation is to divide the slotted channel into a master frame consisting of two subframes. At the beginning of each frame there is a short reservation subframe, during which the users transmit reservation requests on the numbered packet slots that follow. These short, standard-format messages are heard by all other users, who must keep a record of all outstanding reservations. A reservation consists of a slot number, and the number of succeeding frames in which that slot is to be reserved. For example, user K has 6000 bits to transmit, and, according to his tables, slot number 7 has no reservation on it. Therefore, during the next reservation subframe user K would transmit the reservation message "K,7,6," meaning that user K is reserving slot 7 for the next six frames. Each other user hearing the reservation would log the reservation in their local tables and, presumably, would not try to use packet slot 7 during the next six frames. Collisions might occur only if two users attempt to transmit their very short reservation messages at exactly the same time during the reservation subframe.

This reservation technique is extremely efficient for a community of users whose messages are consistently longer than a single packet. Messages of only a single packet (or less) in length are forced to go through the time delay of having to make a reservation for just a single slot in a single frame. On the other hand, successful transmission of that packet on the first attempt is essentially assured.

Any of these reservation or capacity allocation techniques dynamically assign the capacity of a slotted ALOHA channel to active users, thereby achieving higher efficiency than is possible under purely random conditions. Naturally, they also impose some additional overhead, processing, and complexity on the operation of the channel, the terminal equipment, and the users. Under ideal conditions the reservation techniques reach the level of performance achievable with static time-division multiplexing, but they do it so that the resources are adaptable to changing demands of the user community.

COMPARISON OF MULTIPLE ACCESS TECHNIQUES

Table 6-1 summarizes the various multiple access techniques. In an ideal case we could have a single-server system (represented by the **queueing** theory notation **M/D/1,** meaning Poisson arrivals, deterministic packet lengths, and a single server) with sufficient buffering to hold temporary overloads. Such an ideal case can consistently achieve nearly 100% occupancy of the channel, though at the

Table 6-1. **Multiple Access Technique
Summary**

Technique	Examples	Characteristic
Ideal	M/D/1 queue	Statistical multiplexing
Preassigned	TDMA FDMA	Static
Demand access	Polling Reservation	Dynamic
Collision	Pure ALOHA Slotted ALOHA	Uncontrolled

expense of sometimes considerable delay. Preassigned allocation techniques, such as time-division or frequency-division multiple access techniques provide a static allocation of capacity, which can grossly waste capacity if some of the assigned users have little or nothing to transmit over a period of time.

Demand access techniques, such as polling and the reservation methods in the slotted broadcast channel, are highly dynamic and serve a mix of high- and low-capacity users well. However, if we have a community of homogeneous, low-demand users, the demand access techniques are too complex for the benefits gained. Finally, the collision-based techniques—ALOHA and slotted ALOHA— provide a relatively uncontrolled, technically simple mode of access to the broadcast channel.

The characteristics of these access techniques with regard to overhead, empty slots, and collisions are summarized in Table 6-2. The static assignment methods have no operational overhead and no collisions, but when users have nothing to transmit, the capacity is wasted as empty time or frequency slots. Dynamic reservation techniques have some capacity allocated to system overhead, but no capacity is wasted on either empty slots or collisions when demand is high. The uncontrolled, or random, systems have neither overhead nor empty slots when the system demand is great, but they lose capacity to collisions.

Table 6-2. **Characteristics of Resource
Sharing Techniques**

	Overhead	Empty Slots	Collisions
Static assignment	No	Yes	No
Dynamic reservation	Yes	No	No
Uncontrolled	No	No	Yes

SUMMARY

1. The capacity of the purely random ALOHA channel—about 18% of the basic channel bit rate—can be significantly improved by imposing additional discipline and control over user access to the channel.

2. The slotted channel could allow a capacity of 37% to be reached, and permitting one of the users to capture the receiver could raise capacity to as much as 57% of the basic channel bit rate.

3. By adding additional intelligence and mixing users with different average utilization demands, dynamic reservation techniques can often yield nearly 100% utilization of the channel.

SUGGESTED READING

JACOBS, IRWIN MARK, BINDER, RICHARD, and HOVERSTEN, ESTIL V. "General Purpose Packet Satellite Networks." *Proceedings of the IEEE*, vol. 66, no. 11 (November 1978), pp. 1448–1467.

This paper surveys the range of techniques applicable to satellite packet broadcasting for use in a general purpose environment. It introduces and discusses priority-oriented demand assignment (PODA), which schedules satellite channel capacity according to the highly variable needs of different user classes. The paper concludes with the results of experiments using these techniques in actual operation on a trans-Atlantic link as part of the ARPANET.

KLEINROCK, LEONARD, and LAM, SIMON S. "Packet Switching in a Slotted Satellite Channel." National Computer Conference, New York, June 1973. *AFIPS Conference Proceedings*, vol. 42. Montvale, N.J.: AFIPS Press, 1973, pp. 703–710.

This is one of the fundamental papers exploring the slotted channel both in a random mode of access and with various mixes of large- and small-demand users. The paper presents a comprehensive treatment of the various performance tradeoffs, delay/capacity curves, and user loading curves.

ROBERTS, LAWRENCE G. "Dynamic Allocation of Satellite Capacity Through Packet Reservation." National Computer Conference, New York, June 1973. *AFIPS Conference Proceedings*, vol. 42. Montvale, N.J.: AFIPS Press, 1973, pp. 711–716.

Roberts presents a complete treatment of the most flexible of the reservation techniques discussed in this chapter, one using a reservation subframe at the beginning of each master frame. Quantitative and graphical results are derived and presented for a wide range of operating conditions. The paper includes numerous comparisons with the operation of the random channels.

7

Distributed Radio Networks and Carrier Sense Techniques

THIS CHAPTER:

will look at the ways a terrestrial packet broadcast system can operate in a limited geographical region.

will examine the packet interference and collision process that limits the area over which such a system can reasonably operate.

will show how listening to the channel before transmitting can make current channel status more meaningful and thereby improve performance.

In the last few chapters we explored the application of satellites to the creation of broad-coverage packet broadcast networks. There are many different techniques, employing a variety of operational protocols, that allow a very large number of users to be served simultaneously by the limited capacity of the satellite channel. In this chapter, however, we will focus on how the concept of random packet broadcasting can be applied to an entirely terrestrial network. In later chapters, we will see how it is possible to combine local packet broadcasting with long-distance satellite transmission to optimize the cost and design of large distributed communications systems.

TERRESTRIAL TRANSMISSION AND RADIO-BASED NETWORKS

The General Situation

The general situation we will use to understand the difference between the satellite-based operational environment and the terrestrial radio-based system is illustrated in Figure 7-1: a large number of users, each of whom has a relatively low total transmission demand but a potential need to transmit at high data rates

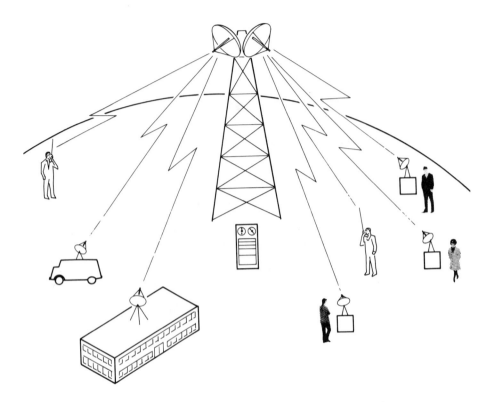

**Figure 7-1. Operational environment of users communicating over a
shared terrestrial radio channel.**

when actively using the system. The users are all located within "line-of-sight"
distance from the central location, which may be a single computer with which
all of the users are trying to interact, or a switch that acts as an entry point into
the larger network covering a broader geographic area. The central location may
even be a satellite earth station, which all the users share to gain entry into one
of the many ALOHA network implementations.

Limitations

If the channel is operating at frequencies in the VHF, UHF, or higher ranges,
the maximum distance between the transmitters and receivers depends upon the
radio line-of-sight, accounting for the curvature of the earth. This distance is
given by the approximate formula:

$$D_{max} = 1.2(\sqrt{H_t} + \sqrt{H_r})$$

where D_{max} is given in miles, and H_t and H_r are the height of the transmitting and
receiving antennas, respectively, in feet above the ground.

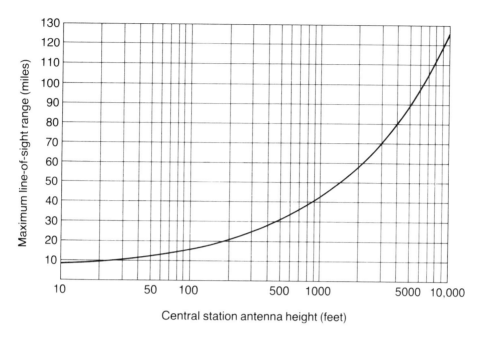

**Figure 7-2.   Plot of maximum line-of-sight range versus height of
central station antenna.**

Figure 7-2 shows the maximum radio range as a function of the height of the central station antenna, assuming that the individual user antennas are approximately 20 feet above the ground (e.g., mounted on mobile vans, small office buildings, utility poles, etc.). This range is, of course, just an estimate, since it depends on local terrain, interfering obstacles, reflections from natural and man-made objects, and atmospheric conditions.

In packet broadcast systems operating with terrestrial radio facilities, line-of-sight distance is not the only limitation to the successful delivery of a subscriber's packets to the central location. The major problem is the decrease in received signal power from a remote transmitting station, which is proportional to the square of the distance between the transmitter and the receiver. This means that, if user B is 4 miles from the receiver, and user A is 2 miles from the receiver, then the receiver will detect user A's signal as four times stronger than user B's (assuming that they are using similar equipment and the same initial transmitter power). Therefore, when two users transmit at the same time using packet broadcasting techniques, the user who is closer to the central station is much more likely to capture the receiver and be received correctly, than the user who is further from the central station. The expected delay results from more distant users having to repeat their transmissions many more times, on the average, before being received successfully. The impact of this phenomenon is a theoretical range beyond which a terrestrial packet broadcasting user cannot expect to use the system successfully.

Gaining Information from the Channel

Another significant aspect of terrestrial packet broadcast systems is the user's ability to gain useful information simply by listening to the common channel. Listening to current information on the satellite channel was of no use in determining whether or not the channel was available since the information was at least a quarter-second old. In the terrestrial system the circumstances are quite different. Though the typical packet (a 1000-bit packet on a 50,000 bits/second channel) may last 20 milliseconds, the propagation time between the transmitters and receivers is on the order of 0.1 millisecond (i.e., 20 miles at a propagation velocity of 186,000 miles per second). In other words, since the packet is about 200 times longer than the typical delay in its arrival at the central location, it is very probable that hearing a currently inactive channel means the channel really is quiet, and a new packet has a good chance of being received without interference. The possibility of listening to the channel before attempting transmission suggests a number of new modes of packet broadcast operation, collectively known as *carrier sense multiple access*, which will be discussed later in this chapter.

SPATIAL CAPACITY OF A TERRESTRIAL ALOHA SYSTEM

The Importance of Location

The impact of the geographical distribution of potential users around the central station in a terrestrial ALOHA system can be analyzed by a procedure analogous to that used for the satellite-based system covered in Chapter 5. However, in this case, the mathematics must also account for the fact that users closer to the central station will, in general, capture the receiver.

This situation is depicted in simplified form in Figure 7-3, which shows distributed users and a centralized station located at the center of the service area. Let us assume n users per square mile—that is, n is the user density—and that each user requires an average throughput of s_o, where s has the same meaning as in the satellite-based cases. Thus:

$$n = \text{user density (users per square mile)}$$
$$s_o = \text{average throughput per user}$$
$$g_o = \text{average traffic per user}$$
$$ns_o = S_o = \text{normalized throughput per square mile}$$
$$ng_o = G_o = \text{normalized traffic per square mile}$$

Note, however, that, although all users have the same average throughput requirement, the amount of traffic generated by each user will depend on his location relative to the central station.

Users close to the central station will generate traffic at a rate just slightly greater than their actual throughput. Because they are close to the central station, very few of their packets will be interfered with, and they will very rarely have to

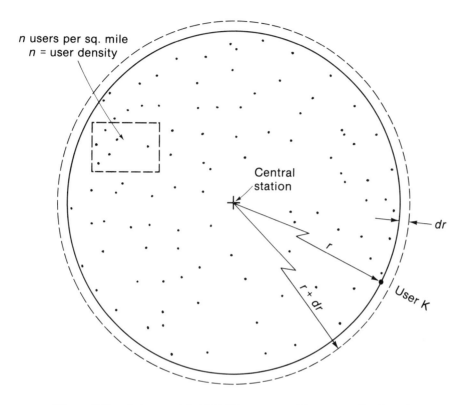

Figure 7-3. A terrestrial ALOHA system with a centralized station.

repeat a packet. Conversely, users far away from the central station will generate much more traffic than their achieved throughput. On the average, each successful packet will have to be repeated many times to break through the barrier of closer stations. Therefore, although S_o, the normalized throughput, can be considered a constant, G_o will be a function of r, the distance of the user from the central point. The normalized system traffic can thus be represented as $G_o(r)$, indicating that channel traffic generation is dependent on the distance from the central station.

To make this problem mathematically tractable, we will also assume a situation of "perfect" capture, which means that, as long as a given user is even slightly stronger than an interfering station, the stronger traffic will be received perfectly, and the weaker traffic will be lost. In reality, an appreciable difference in strength is really required to create the capture effect.

Estimating Channel Performance

Now let us return to Figure 7-3 and focus on user K, who is located at an arbitrary distance, r, from the central station.

If user K transmits a packet, it will be received successfully only if no other closer user transmits at the same time. The total packet arrival rate for users closer to the central station than user K is given as:

$$G = 2\pi \int_o^r G_o(r)\, r\, dr$$

which is the area integral of all users located within the distance r from the central station.

Since the user traffic generated within the distance r forms a Poisson process, the probability of user K being successful is:

$$P_k = e^{-2G} = e^{-2\int_o^r 2\pi G_o(r)r\, dr}$$

$$P_k = e^{-4\pi\int_o^r G_o(r)r\, dr}$$

However, in a "narrow ring" of width dr, at a distance r from the central station, the incremental throughput must be equal to the incremental traffic multiplied by the probability of success at that distance. The incremental throughput is the area of the ring times the throughput density, the incremental traffic is the area of the ring times the traffic density, and the success probability is given above. Thus:

$$2\pi r S_o\, dr = 2\pi r G_o(r)\, dr\,(P_k(r))$$

$$2\pi r S_o\, dr = 2\pi r G_o(r)\, dr\, e^{-4\pi\int_o^r G_o(r)r\, dr}$$

$$S_o = G_o(r) e^{-4\pi\int_o^r G_o(r)r\, dr}$$

The last expression does not help to define channel operation clearly. However, if each side of the expression is differentiated with respect to r, and each time the exponential appears it is replaced with $S_o/G_o(r)$, a more meaningful form results. If the traffic throughput is constant, then its derivative with respect to r is zero, and the following results:

$$O = G_o(r)e^{-4\pi\int_o^r G_o(r)r\, dr}(-4\pi r G_o(r)) + G'(r)e^{-4\pi\int_o^r G_o(r)r\, dr}$$

$$\frac{G_o(r)S_o}{G_o(r)} 4\pi r G_o(r) = \frac{G'(r)S_o}{G_o(r)}$$

or

$$4\pi r G_o{}^2(r) = G'(r)$$

Solving the Differential Equation

This result is a relatively simple differential equation, subject to the constraint or boundary condition that, at $r = 0$, $G_o = S_o$, since, at the center of the service area there is no interference, and the throughput and traffic have to be exactly equal.

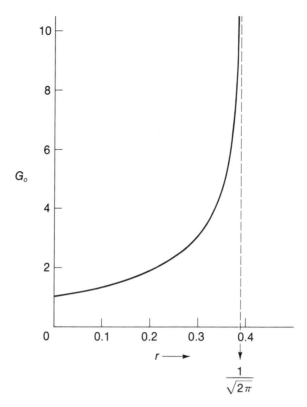

Computed for $S_o = 1$

In general, Sisyphus distance = $\dfrac{1}{\sqrt{2\pi S_o}}$

Figure 7-4.   Plot of the relationship between average traffic generated and distance from the central station.

The solution to the differential equation yields the desired relationship between the average throughput and the distance from the central station, which is given as:

$$G_o(r) = \frac{S_o}{1 - 2\pi r^2 S_o}$$

This relationship is plotted in Figure 7-4. We can see that, when r reaches a value of $1/\sqrt{2\pi S_o}$, the value of G_o tends toward infinity. In other words, the distance to the central station is so great that, no matter how often a user transmits (that is, no matter how large his traffic becomes), some user is always closer, and therefore stronger, interfering with him and thereby preventing him from achieving any successful throughput.

This maximum distance, R_o, given simply as $1/\sqrt{2\pi S_o}$, has been called the **Sisyphus distance,** after the evil Greek king who was condemned to an eternity of rolling a huge stone up a hill, only to have it roll back again just before reaching the top. Note that, if the Sisyphus distance is R_o, the total throughput within the circle to this distance is simply the area times the throughput density, but the maximum value is one-half. That is,

$$R_o = 1/\sqrt{2\pi S_o}$$

$$\text{Total throughput} = \pi R_o{}^2 S_o = \frac{\pi S_o}{2\pi S_o} = \tfrac{1}{2}$$

This simply means that the total normalized throughput of the channel, on a spatial basis, is one-half the channel capacity. Theoretically, users can be accommodated out to the Sisyphus distance, which is a function of the normalized throughput density.

Examples of Terrestrial Packet Radio System Limitations

In order to illustrate the spatial limitations on a terrestrial packet radio channel, we will look at two illustrative examples.

A High-Capacity Channel. The first example uses a high-capacity channel, operating at a basic rate of 1,500,000 bits per second, with a packet size of 1000 bits. The channel thus has a capacity of 1500 packets per second. It is desired to support users out to a distance of 20 miles. The average user generates a full packet every 100 seconds, resulting in a throughput requirement of 0.01 packet per second, per user. Thus:

$$R_o = 20 \text{ miles}$$
$$C_o = \text{channel capacity} = 1500 \text{ packets/second}$$
$$s = \text{individual user throughput} = 0.01 \text{ packet/sec}$$

With this information we can compute the allowable user density and the maximum number of users the channel can support.

From the Sisyphus relationship we know that:

$$S_o = \text{normalized throughput} = \frac{1}{2\pi R_o{}^2}$$

which, for a distance of 20 miles, becomes:

$$S_o = \frac{1}{2\pi(20)^2} = \frac{1}{800\pi} = 4 \times 10^{-4}$$

However, S_o is defined as the normalized traffic throughput, which is the packet throughput density (per square mile) divided by the channel capacity. Thus:

$$S_o = \frac{\text{packet throughput rate} \times \text{users per sq mile}}{\text{channel capacity}}$$

but this has to be equal to the value for S_o computed from the Sisyphus relationship. Therefore, equating the two representations of S_o yields:

$$S_o = 4 \times 10^{-4} = \frac{0.01 \times n}{1500}$$

where n is the number of users per square mile. Solving for n yields:

$$n = \frac{1500(4 \times 10^{-4})}{0.01} = 60 \text{ users/square mile}$$

The total number of users in the system is simply the user density times the total area within the Sisyphus distance, or:

$$N = \pi R_o^2 n = \pi 20^2 \times 60 = 75,000 \text{ users}$$

Note that, if there are 75,000 users, each generating 0.01 packet per second of throughput, the average throughput is 750 packets per second for the total channel, or, with 1000-bit packets, 750,000 bits per second. As the theory predicted, this is one-half the total gross channel bit rate of 1,500,000 bits per second.

A Smaller Channel.   A second example will scale the problem down to a much smaller channel and determine the maximum range to which users can be supported.

Let us suppose that user density is two users per square mile, the channel rate is 9600 bits per second, or 9.6 packets per second for 1000-bit packets, and that users generate one packet every 100 seconds, or 0.01 packet per second. The normalized throughput density, S_o, is the ratio of the aggregate user throughput density to channel capacity, so that:

$$S_o = \frac{\text{throughput density}}{\text{channel capacity}}$$

$$= \frac{2 \text{ users/sq mile} \times 0.01 \text{ pac/sec}}{9.6 \text{ packets/sec/user}}$$

$$S_o = 0.0021$$

But S_o is limited by the Sisyphus relationship, so that:

$$S_o = 0.0021 = \frac{1}{2\pi R_o^2}$$

$$R_o^2 = \frac{1}{2\pi(0.0021)} = 75.8$$

or

$$R_o = 8.7 \text{ miles}$$

The maximum number of users is the user density times the service area, or $2 \times \pi R_o^2 = 2\pi(8.7)^2 = 480$ users. Each user is generating packets at a rate of 0.01 packet per second, yielding an overall generation rate of 480×0.01, or 4.8 packets per second, which indeed is one-half the channel capacity of 9600 bits per second.

This example shows that, overall, the channel with relatively low total capacity of 9600 bits per second is capable of supporting 480 uniformly distributed users, over an area out to a distance of 8.7 miles from the central station. Interestingly, the central station antenna would have to be only about 25 feet high to insure line-of-sight range to this distance.

Improving the Packet Radio Channel

Our description of the limitations on the terrestrial packet radio channel is based on many idealizing assumptions. However, we can reconsider the impact of some of these assumptions, particularly the effect of perfect capture. Furthermore, many practical modifications can be made to overcome limitations. The limitation of the Sisyphus distance could easily be overcome, for example, by providing users further from the central station with more transmitter power. In fact, if the transmitter power were increased as the square of the distance to the central station, the traffic density would remain constant with distance, and no Sisyphus effect would be observed. Of course, eventually we would reach a distance beyond which it would be neither practical nor possible to increase the transmitter power. Further thought about the problem leads to consideration of the next approach, which employs the ability to sense the channel occupancy before transmitting.

CARRIER SENSE MULTIPLE ACCESS—
LISTEN BEFORE YOU SEND

A terrestrial radio system offers a distributed community of packet users the ability to listen to the channel before transmitting, thereby eliminating most of the problem of overlapped packets. The so-called carrier sense multiple access (**CSMA**) techniques have been the object of a great deal of theoretical study, with many different algorithms and possible modes of operation evolving. In addition, the combination of the carrier sense principles and the concepts of packet reservation can create very powerful and flexible systems, capable of operating very close to 100% of capacity over a broad range of conditions.

Operational Principles and Problems

As the name *carrier sense multiple access* implies, there is some way for users to determine if some other user is currently transmitting a packet. Since, in general, individual users can communicate with the central station but not directly with each other, this sensing ability may require the central station to transmit a special signal, on some other frequency, to indicate when the users' input channel is occupied. Nevertheless, even when users' terminals are able to sense the operational state of the channel, there are still some uncertainties in the operation of the channel.

Although the transmission delay between the typical user and the central station is short, especially compared to the round-trip delay to a satellite, the propagation delay is not negligible. When the packets are relatively short or the channel rate is very high, the transmission time of the packets may be an appreciable percentage of the propagation time to the central station.

For example, with a channel rate of 1,500,000 bits per second, a 1000-bit packet will have a packet length of about 0.67 millisecond. The propagation time to a user 60 miles from the central station would be about one-half this value. This means that, if a user started transmitting from a distance of 60 miles from the central station, half his packet would already be transmitted before another user could detect that his transmission had begun. Thus, any user who was closer to the central station and thus capable of interfering with the user 60 miles distant might conclude, even after sensing the channel, that the channel was vacant.

In order to analyze this problem, we will use a parameter, often designated a, which is simply the ratio of the maximum propagation delay to the packet length. In the example just mentioned, the value of a would be 0.5, indicating that the propagation delay to the most distant users in the system would be one-half the packet length. For terrestrial systems a generally has a small value, most likely less than one. For satellite systems, on the other hand, a is a larger number, typically 10, 50, 100, or more, because of the minimum 250-millisecond round-trip delay to the satellite.

User Persistence

Another key characteristic of CSMA techniques—user persistence, or simply **persistence**—concerns user behavior when the user senses the channel and finds it busy. The relevant operating algorithms are classed as either persistent or nonpersistent.

The nonpersistent algorithm works as follows. Upon receiving a transmit command from the user, a user terminal device senses the channel. If the channel is sensed as vacant, the terminal transmits. There is a good probability, especially if a is very small, that the packet will succeed in getting to the central station. However, if the terminal senses the channel to be occupied, the nonpersistent user sets a random timer and, after the timer elapses, senses the channel again. The process is repeated until the channel is sensed to be vacant, and the terminal then transmits the packet.

The persistent algorithm works quite differently. The terminal senses the channel, and if it finds it vacant, of course, transmits the packet. However, if the persistent terminal senses the channel occupied, it continues to sense the channel.

The instant the persistent terminal senses the channel vacant, it begins to transmit. This can achieve very high channel occupancy since the channel is not permitted to go idle before the next persistent user begins transmitting. A problem occurs when two or more persistent users have packets to transmit, each one sensing the channel and waiting to pounce on it. As soon as the channel becomes idle, all persistent users who have been waiting begin to transmit simultaneously, causing immediate collisions and thus reduced traffic throughput.

The persistent user algorithm can be improved by the addition of one parameter. Each user senses the channel persistently. However, when the channel becomes idle, the users do not always transmit immediately. In fact, the users transmit with some probability—p (where p lies between zero and unity).

For example, let us assume two users and p equal to one-half. When the channel becomes idle, each user in effect flips a coin. If the coin comes up heads, one user—say, user A—transmits, and if it comes up tails, he does not transmit. If user A does not transmit, he waits a time sufficient to account for the maximum propagation delay and then reinitiates normal carrier sense operation. By this time user A would begin to sense the existence of any other user's transmission. Given two users and p equal to 0.5, the following situation would result: One-fourth of the time user A would transmit, and user B would wait; one-fourth of the time user B would transmit, and user A would wait; one-fourth of the time both users would transmit; and one-fourth of the time both users would wait. The probability of a collision is reduced from unity for the fully persistent case to one-fourth in the so-called $\frac{1}{2}$-persistent case.

Persistent algorithms are defined by the nomenclature of p-persistent, where p is the transmission probability at the instant the channel becomes idle. The original persistent algorithm—that is, where the users always transmit as soon as the channel becomes idle—would thus be termed 1-persistent since the users transmit with the probability of unity on hearing the busy channel become idle.

Mathematical Analysis

As might be expected, the mathematical analysis of the persistent and non-persistent CSMA techniques is quite complex. The derivations can be found in the references at the end of this chapter; we will describe only the more important results below to illustrate the potential channel performance using these techniques.

The 1-Persistent User

Figure 7-5 shows the throughput curves for a 1-persistent channel for various values ranging from $a = 0.0$ to $a = 1.0$. Notice that a value of $a = 0.0$ means that the propagation delay is truly negligible compared to the packet length. Since this is a theoretical impossibility, such a small value of a really implies that the system is operating either over a very small geographic area (possibly a mile or less) or

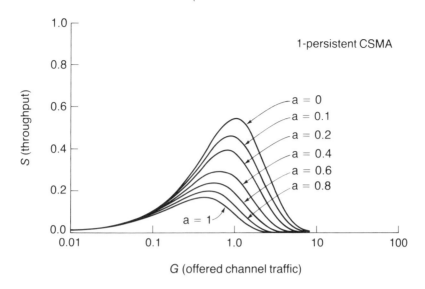

**Figure 7-5.   Channel throughput versus channel traffic: plot for
1-persistent case and various values of propagation delay.
© 1975, IEEE.**

with very long packets (for example 1000-bit packets on 9600-bits per second
channels, resulting in 100 ms packet lengths). In any case, very small values of a
mean that the users have very accurate information about the channel when they
listen to it. This assumption naturally leads to the highest possible channel throughput for the 1-persistent operation.

The p-Persistent User

Figure 7-6 shows the relationship of the channel throughput and traffic for a
p-persistent channel, with the values of p varying from 0.01 to 0.99. In each case
the value of a, the ratio of propagation delay to packet length, was held constant
at 0.05. It is interesting that, at a value of p between 0.01 and 0.20, the maximum
channel capacity changes very little. As p is increased, the traffic corresponding to
the maximum throughput does shift from higher to lower values. Recall that the
ratio of traffic to throughput is a measure of average delay since it is a direct
measure of the average number of times a given packet has to be repeated before
it is received successfully. This effect clearly indicates the existence of an optimal
value of p—that is, the maximum value of channel throughput at the lowest
possible value of average delay. Figure 7-6 further shows that the channel capacity
is decreased, without significant improvement in delay, as the value of p is increased
above a value of $p = 0.20$.

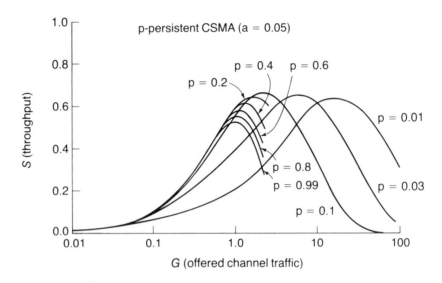

Figure 7-6. Channel throughput versus channel traffic: plot for propagation delay of 5% of packet length and various values of transmit probability when channel becomes clear. © 1975, IEEE.

Comparison of Various Channel Protocols

Figure 7-7 summarizes various algorithms, including the basic ALOHA, slotted ALOHA, and several carrier sense situations. The CSMA cases are all computed at the value of *a* equal to 0.01. The significant increase in channel capacity, albeit for increased values of expected delay, can readily be observed on this figure. It is interesting to note that the overall channel capacity is as great for the nonpersistent algorithms as for any of the other algorithms, although the average delay is longer.

The impact of propagation delay is illustrated in Figure 7-8. This figure plots the maximum channel capacity as a function of the value of *a*, ranging from $a = 0.001$ to $a = 1.0$. The value plotted is, in effect, the peak value obtained on the kind of curves shown in the previous figures. The point here is that, as the value of *a* reaches the vicinity of 0.2, the carrier sense process fails to yield any improvement compared to the basic nonsensing ALOHA techniques because the information sensed on the channel is too "old." This is important to remember in considering a carrier sense operation with fairly high-capacity channels since, if the basic bit rate is high, the packet durations will be short, leading to large values of *a*.

Finally, Figure 7-9 shows the throughput-delay tradeoff for various algorithms, all computed at a value of $a = 0.01$. For any given value of delay (in Figure 7-9 normalized to the packet length), the overall channel throughput can

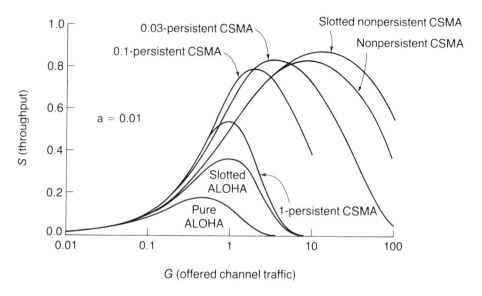

Figure 7-7. Channel throughput versus channel traffic: Delay of channel is assumed to be 1% of the length of a packet. © 1975, IEEE.

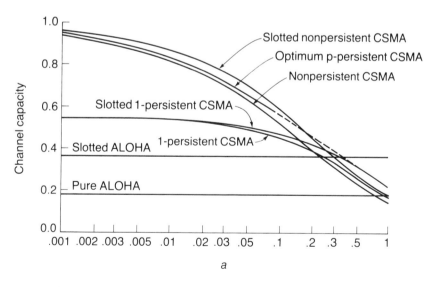

Figure 7-8. Effect of propagation delay on maximum channel capacity. © 1975, IEEE.

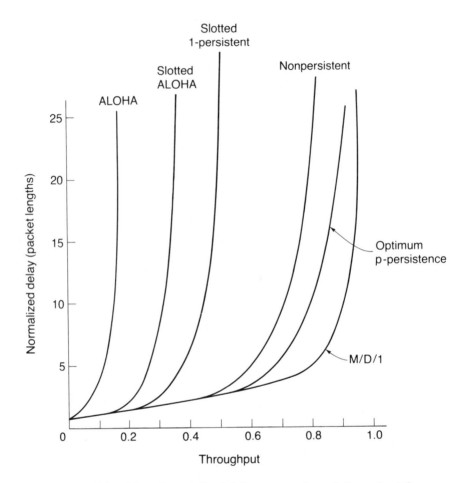

Figure 7-9. Plot of normalized delay versus channel throughput for various network access techniques. ($a = 0.01$)

be substantially increased by the proper channel operational algorithm. Remember, however, that a value of $a = 0.01$ would correspond roughly to a 30-mile system, operating at 50,000 bits per second with 1000-bit packets. As the value of a increases, the advantages of the carrier sense systems begin to fade away rapidly as the information sensed by the users becomes increasingly stale. This consideration will be important in the next chapter when we consider local networks that utilize cable connectivity and very high data rates.

SUMMARY

1. The basic satellite broadcast concept of a common radio frequency channel, with all members of the network being able to freely transmit on the channel, can be carried over directly to terrestrial networks.

2. The fact that the terrestrial environment involves much shorter transmission distances enables users to use information in the channel, in real time, to affect and modify their behavior.

3. Furthermore, users located further from the central station or system repeater operate at a serious disadvantage compared to stations located closer. More distant signals are generally dominated by the closer stations, making successful transmission difficult without many retransmissions. The Sisyphus distance is the theoretical point beyond which a terrestrial user cannot reach the central station successfully.

4. A number of different algorithms make it possible for users to sense the current state of the channel before deciding whether or not to transmit. These include both persistent and nonpersistent carrier sense multiple access (CSMA).

5. If the propagation delay is short relative to the packet length (approximately 20% or less), the information that the users hear by monitoring the channel is sufficiently "up-to-date" to permit a useful decision and significantly enhance channel performance. However, if the information is old, the basic ALOHA or slotted ALOHA techniques provide service that is as good, or better than, CSMA techniques.

SUGGESTED READING

The technology of terrestrial packet radio networks is rich in recent literature, both theoretical and practical. Only a small portion of the technology could be covered in this chapter. The suggested references provide a detailed coverage of the total technology of terrestrial packet broadcasting, including an extensive experimental system developed under contract to the Advanced Research Projects Agency (ARPA) of the U.S. Government.

ABRAMSON, NORMAN. "The Throughput of Packet Broadcasting Channels." *IEEE Transactions on Communications*, vol. COM-25, no. 1 (January 1977), pp. 117–128.

This paper provides a consolidated treatment of packet broadcasting. Particularly relevant to this chapter is the derivation of the spatial capacity of the terrestrial networks and the limitations of the ability of distant users to reach the central station.

KAHN, ROBERT E., GRONEMEYER, STEVEN A., BURCHFIEL, JERRY, and KUNZELMAN, RONALD C. "Advances in Packet Radio Technology." *Proceedings of the IEEE*, vol. 66, no. 11 (November 1978), pp. 1468–1496.

This comprehensive paper summarizes the practical aspects of packet radio technology. Included are discussions of resource organization, control of the network, and application of a spread spectrum transmission technology to achieve high transmission bandwidths with low incidence of interference. Of particular interest are the descriptions of experimental operational equipment that was deployed and tested in the San Francisco Bay Area.

KLEINROCK, LEONARD, and TOBAGI, FOUAD A. "Packet Switching in Radio Channels: Part I and Part II." *IEEE Transactions on Communications,* vol. 23, no. 12 (December 1975), pp. 1400–1433.

The source of many of the figures used in this chapter, this two-part paper reports the research on the operation and performance characteristics of carrier sense multiple access. The basic premise is that all users can directly hear the transmissions of all the other users, or the central station provides an indication of when the channel is busy. Much of the technology of packet radio systems is based on the results of this research.

METCALFE, ROBERT M., and BOGGS, DAVID R. "Ethernet: Distributed Packet Switching for Local Computer Networks." *Communications of the ACM,* vol. 19, no. 7 (July 1976), pp. 395–404.

This is an early paper in the history of the Ethernet, Xerox Corporation's broadly applied local area network system using a passive broadcast medium with no central control. Switching is achieved using packet address recognition. The paper explains the basic design principles and a model for estimating the performance of the network under heavy loads.

8

Local Networks—Applying the Principles of Distributed Communications

---

THIS CHAPTER:

will show how the relatively low cost of capacity over short distances has led to the modern electronic office environment.

will place the concept of local networks within the framework of larger national and worldwide networks.

will explore the economic and design advantages of local networking.

THE ADVENT OF THE ELECTRONIC OFFICE

In the late 1970s a number of factors came together to create a revolutionary new marketplace. The most important of these factors was the rapid progress made in large-scale integrated circuits (**LSI**) and the embodiment of LSI circuits into single-board, and later single-chip, microcomputers. These microcomputers resulted in the application of intelligent, computer-driven controls and features to everyday devices. By 1978 everything from typewriters to automotive ignition systems, from copying machines to electronic ovens, from televisions sets to children's toys featured microcomputers or microprocessors in their design and operation. At the same time, the availability of more general purpose small, inexpensive computers moved many data processing functions "out of the back room and into the office." Word processors displaced typewriters, electronic mail displaced office memos as well as many telephone calls, and improved voice and video communications displaced many face-to-face meetings and business trips. Video cathode ray tube (**CRT**) terminals permitted immediate, on-line entry of data directly into larger and more powerful data processing facilities without the need for intermediary data translation or key-punching functions.

The net result of all of these changes was the creation of the so-called automated, or electronic, office. A wide variety of functions in modern offices are accomplished directly or are assisted by computerlike equipment. Thus, in the course of the day an executive secretary might use an electronic typewriter, a word processor, a document reproduction machine, a document distribution system, an electronic mail system, and a computer-based calendar and scheduling system. A sales executive might make use of an electronic mail system, a computer data base for statistical information retrieval, a public information system for market information, and a teleconference system.

The processor-aided functions needed by these two individuals, and many others in the same office, would likely be embodied in many different physical devices. Tying the various processor-aided functions together greatly streamlines the work flow and enhances the ability to move, manage, control, and use information. Though at first such integration required use of common equipment, or equipment designed as a system by a single manufacturer, the transition to local network facilities, which permit very flexible, high-capacity communication and interconnection, has greatly enhanced the ability to integrate many office automation functions.

Reversing the Traditional Communications Tradeoffs

The basic premise embodied in the first seven chapters of this book, whether we were dealing with satellite or terrestrially based networks, was that the communications capabilities were relatively expensive resources. In order to use the communications transmission facilities efficiently, processing power was applied to the basic channels, and the raw capacity of those channels could then be allocated on a demand basis to users with current information to be transmitted. Thus, packet switches and shared channels using ALOHA, capacity reservation, and a wide variety of networking methods were very effective communications techniques.

Figure 8-1 shows the cost trends for computing and communications using packet switching. The curves in this figure, based on ARPANET experience, represent the rough costs for processing 1 million bits of data and for transmitting 1 million bits of data across the United States. In 1971, the cost of processing a million bits of data and the cost of transmitting a million bits were about equal, at approximately 35¢. However, as computers grew less expensive, by the late 1970s the cost of transmission was at least ten times that of computation on an incremental cost basis. Without some major breakthrough in transmission costs, this differential will be about 100 times by the mid-1980s.

If we look at communications costs on an officewide, rather than a nationwide, dimension, the situation changes dramatically. Extrapolating from Figure 8-1, we arrive at 1980 processing costs per megabit of roughly $0.003, or three-tenths of a cent. If, for example, we can connect all of our office automation machines together with about $25,000 worth of cable and connectors, use the system about 8 hours a day for a 3-year lifetime, and put about 3 million bits per second through the

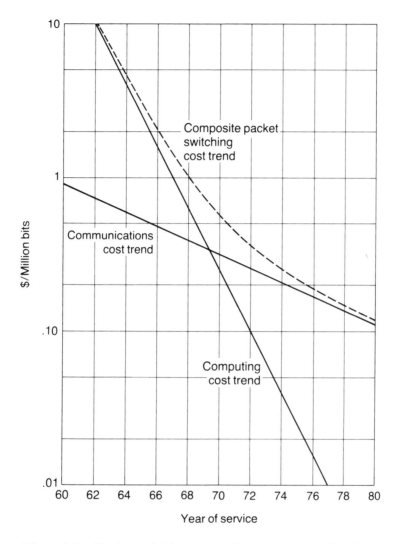

Figure 8-1.   Packet switching cost performance trends. Cost for processing and transmitting 1 million bits of data across the United States. © 1974, IEEE.

local cable, the allocated, incremental communications cost would likewise be about $0.003 per megabit. Thus, within the automated office environment, the cost of transmission is once again close to, or possibly even less than, the cost of processing. In other words, in local networks used over a very limited geographical area, the advantages of processing over communications are less evident. Consequently, using transmission media efficiently without overburdening the processors with a great deal of overhead requires a balanced approach. Many of the techniques that were described earlier, including carrier sense multiple access and capacity reservation, meet this requirement well. These approaches permit efficient channel utilization, a large degree of network structure and interface independence, and a relatively low demand on the complexity of the communications and interface processors.

PLACING LOCAL NETWORKS IN A LONG-DISTANCE ENVIRONMENT

Features and Applications of Local Networks

Local networks can play a very important role in efficiently allowing processor-based devices to utilize nationwide, long-distance communications facilities. At the boundary between local and long-distance networks, various techniques not only achieve efficient transmission of the user data, they also permit continued operation of the nationwide facilities under anomalous conditions or system failures, and fairly allocate the network resources to the various users. National and international standards, such as the Consultative Committee of International Telephone and Telegraphy (CCITT) number X.25, provide the features and protocols necessary for utilization of the national and international networks.

Such varied techniques permit communication with similar or dissimilar equipment over broad geographic distances. As shown in Figure 8-2, user end devices are tied together by relatively simple, low-cost local networks, using interface protocols that minimize the complexity of the local network interface. One device on the local network, the **internetwork gateway,** concentrates the interfacing of the long-distance networks in a single device. The internetwork gateway can accept data from any of the locally connected devices; add the long-distance network protocols, overhead, and controls; and transmit efficiently through the public switched networks to similar distant internetwork gateways. After receiving the information at the distant end, the internetwork gateway converts it back into the simple local network protocol (which may not even be the same as that in the local network at the originating end) for delivery to the destination device.

Local networks employing internetwork gateways reduce the problem of potentially interfacing a large number of devices with a large number of different networks and devices to interfacing many different devices with a single gateway. Figures 8-3 and 8-4 show how the problem of interfacing five computers among three different service networks is greatly simplified by using a local network to achieve the gateways. In this example the word *computer* is used generically; the devices could be any of many automation and information resource devices that may be found in a modern office or industrial situation.

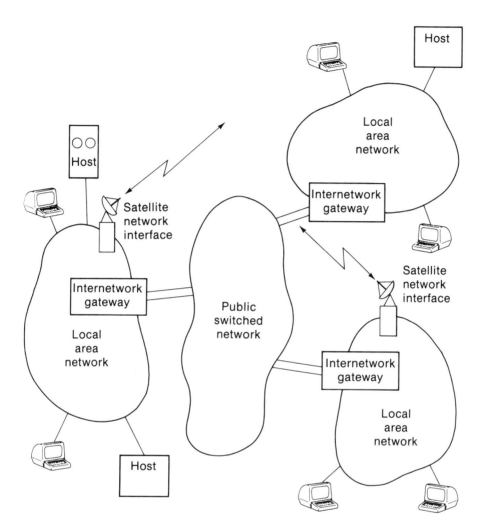

Figure 8-2.   Local networks in a long-distance environment.

Viewed from the user's perspective, local networks are the first in a series of capabilities that permit the interchange of data and information among a wide variety of equipment meeting different applications and functions. The automated office is a contemporary, highly visible, and widely promoted target of local networking. However, local networking can be cost effective and practical in many situations where other solutions and approaches had traditionally been used. In Chapter 9 we will discuss the implementing technology by which local networks accomplish the distributed control of network interface and capacity allocation,

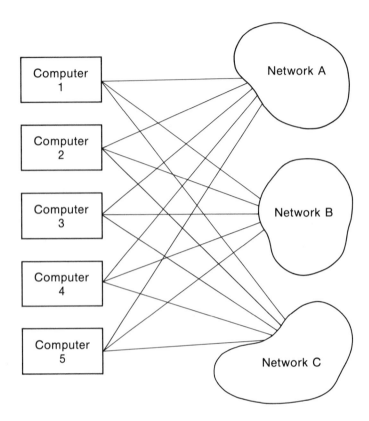

Figure 8-3.   Interfacing without a local network.

and the ability of user information to enter and leave the network largely independent of any centralized resource managers. Coupled with satellites to provide the large-capacity, long-distance communications resources, local networks provide the mechanism for low-cost, widely distributed communications.

Economic Incentives

To understand the economic motivation for using local networks, let us return to Figure 8-1. The economic tradeoff situation illustrated there was limited by the fact that the extrapolated trends were based on the technology of the early 1970s. Figure 2-1 (Chapter 2), which highlights the specific technology of satellite communications on the basis of voice-channel costs, showed that by the early 1980s local connection to satellite facilities had become the dominant part of satellite system cost. However, the possibility of direct interfacing by bringing satellite facilities to the specific locations of the local networks removes much of

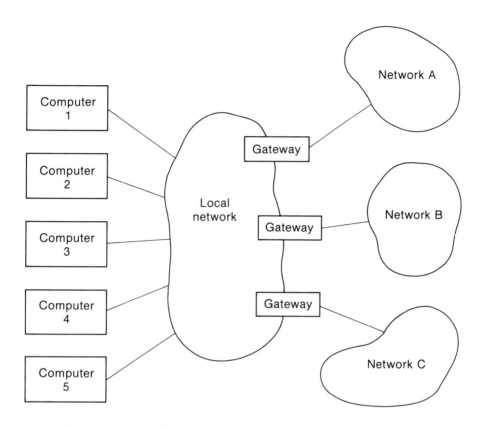

Figure 8-4.   Interfacing using a local network to achieve gateways.

this cost and capacity limitation. Such a case results in a modified version of Figure 8-1.

Figure 8-5 shows the total network cost trend with and without local network influence. It is assumed that, without a local network, individual user devices would have to interface with satellite facilities using shared earth stations. For most users this would necessitate relatively expensive, short-distance, carrier-provided terrestrial interconnections. On a per-voice channel, or more significantly on a per-megabit of capacity, basis these facilities are much more expensive than the satellite capacity itself. Tying many users together through local networks permits the use of an earth station dedicated to the total requirements of the local network, thereby eliminating the need for costly carrier-provided short-distance communications to a shared earth station. These two situations—satellite interface without and with local network—are shown in Figures 8-6 and 8-7, respectively.

Even without the interface to a satellite facility, the economic advantage of using a local network to provide a single interface to one or more public networks for a large number of users can be substantial. Figures 8-2 and 8-4 illustrated this

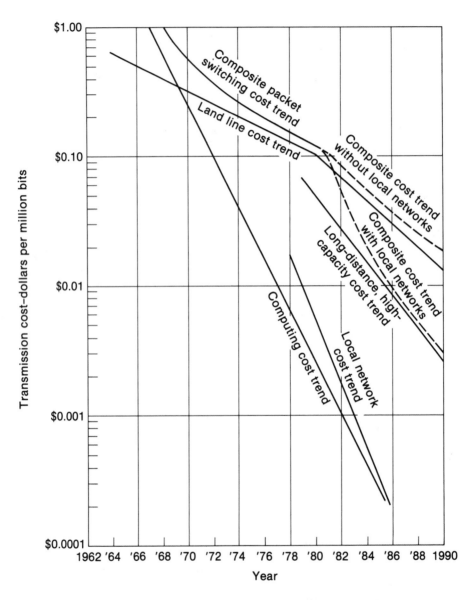

Figure 8-5.   Total network cost trend (projected to 1990).

situation from the perspective of technical and software complexity. In addition to simplifying the interface and protocol complexities for the users, the internetwork gateway, by aggregating demand, can use much higher capacity facilities to tie the users to the nationwide networks. Not only are the higher capacity facilities less expensive on a per-unit capacity basis, but they can be used much more

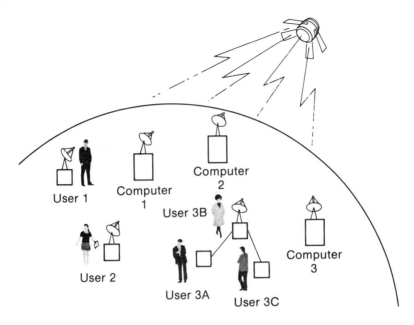

**Figure 8-6.   Satellite interface using multiple earth stations plus
short-haul leased circuits.**

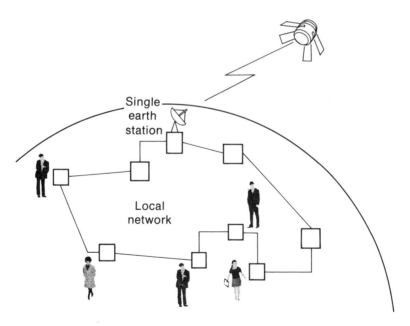

**Figure 8-7.   Satellite interface with a local network using a single
earth station.**

efficiently than the same total capacity divided among a number of smaller facilities.

Design and Operational Incentives

Protocol Simplicity.   The relative simplicity of the interface protocols is a highly desirable feature of local networks. Software has largely displaced hardware as the dominant cost, as well as the source of problems, of data processing systems. Therefore, it is desirable to make the system protocols and controls as simple as possible. Local networks use very simple protocols that can be easily developed and integrated into the software operating systems of user devices. Because of their simplicity, the protocols can be analyzed to insure that they will perform as designed, even in unpredictable network conditions.

Analysis and Design Simplicity.   Because in long-distance networks communications facilities costs dominate processing costs, there is great concern for the total efficiency of the network design. Many tradeoffs have to be analyzed, requiring accurate estimates of total user demand, location of both user and network facilities, total number of both short-term and long-term users to be served, the balance between network-owned versus leased services, the balance between satellite and terrestrial services, and so forth. By contrast, local networks, by largely replacing protocol complexity with low-cost capacity, permit the user devices, as well as configuration of the network itself, to be implemented with a minimum of analysis and detailed design. With logical and intelligent choices of implementing technology based on the general characteristics of the local network users, the network design can be highly dynamic and adaptable to the changing and evolving needs of the user community. The complex issues of overhead structure and overall efficiency are then delegated to the internetwork gateways.

Simple Flow Control.   Nationwide networks must be protected against saturation and overload by a variety of flow-control mechanisms. The flow control operates in conjunction with the interface protocols to limit the amount of data that a given.user can enter into the network at any time. When the capacity of the network is approached, either because many users are active simultaneously or because some of the network resources may be temporarily unavailable, the flow-control mechanisms will reduce the allowable entry rate of information into the networks. The flow control mechanisms interact with the user application software as well as the network interface protocols by establishing variable timing for the entry and exit of both basic information and error protection acknowledgements. Through their large capacity, high speed, and low delay, local networks eliminate or minimize the need for flow control, making interface of the user devices to local networks much easier than interface to long-distance networks.

Simple Interfaces and Hardware Independence.   Local network implementations are able to employ simple interfaces and a large degree of hardware independence.

Again, because of the high capacity available in the local network environment, the network interface looks largely like a high-speed interface between the processing elements and the network, similar to the typical interface between users and their peripheral devices. In addition, the limited number of interface combinations permits the design of standardized interface converters or adapters so that the hardware of many different manufacturers can be interfaced to the common local network. For example, when it established the **Ethernet** structure, Xerox Corporation broadly publicized and openly licensed the design of the Ethernet software so many manufacturers could design their hardware interfaces to be directly attachable to the network controllers.

High Speed and Low Delay.   For the end user, the most satisfactory feature of local networks is the high speed and low delay achieved by the network. Low delay is most apparent in the transparency of the network, where information exchange among the devices is so rapid that any information within the local network moves as rapidly as if there were a direct connection between the various user devices. For example, finished pages from a word processor would be transmitted to a shared page printer as rapidly over the local network as they would if the shared printer were connected directly to the word processor. This feature permits the sharing of many expensive peripheral devices, such as high-quality printers, large-capacity disk storage devices, voice message storage devices, and so forth, with essentially transparent operation for the end users.

Local Network Applications

The combination of features and capabilities of local networks extends to many possible applications beyond those mentioned earlier. Resource sharing of very high-capacity, large-scale processors is used in many scientific and data-based management systems, such as air-traffic control, spacecraft simulation, or nuclear energy research, where a number of powerful processors and peripheral devices are tied together in a localized geographic area. Other important applications of local networks include local distribution of voice and data traffic for long-distance networks and integration of voice and data communication facilities. Local networks provide a mechanism whereby information security and protection devices can be economically applied to users' networks. Because of the cost of the devices themselves, as well as the additional complexity in network administration, information security is still very expensive. However, the lack of proper security can be even more expensive in terms of potential fraud, loss of services, or loss of privacy (e.g., industrial espionage). Local networks can operate safely in protected environments and then use shared protection devices at the internetwork gateways.

Local networks will undoubtedly expand their importance and applications as computers and computerlike information retrieval devices make greater inroads in the home and consumer markets. Systems like videotext and teletext, which combine data distribution with television systems and receivers, and full-scale home computers make the efficient local distribution of digital data using local networks increasingly likely. Recent trends among the telephone companies, including the American Telephone and Telegraph Company (AT&T), to provide data services directly to end-user premises is a step in this direction. Both intelligent services (such as AT&T's Advanced Communications Service, **ACS**) and transport services, using either direct digital service at 56,000 bits per second or X.25 packet switching, are becoming available. Local networks will increasingly be used to improve the cost effectiveness of such consumer data communications services.

SUMMARY

1. The development of computer technology since the early 1950s, along with a change in the balance of transmission versus processing costs, had led to the creation of the electronic office environment.

2. Local networks are very important elements in efficient nationwide long-distance communications systems.

3. Local networks combine cost advantages, design simplicity, and operational efficiency.

SUGGESTED READING

CLARK, DAVID D., POGRAN, KENNETH T., and REED, DAVID P. "An Introduction to Local Area Networks." *Proceedings of the IEEE*, vol. 66, no. 11 (November 1978), pp. 1497–1517.

This article provides a comprehensive technical introduction to local networks, emphasizing the hardware and protocols needed to realize economical, high-speed local communications networks. It contrasts some of the procedures required for effective local operation to the similar functions in long-haul networks, and also considers the interconnection of local networks and long-haul networks.

SIDERIS, GEORGE, and RILEY, WALLACE B. "Local Networks Mushroom to Increase Productivity." *Electronic Design*, vol. 29, no. 20 (September 30, 1981), pp. SS-7 to SS-56.

This article is actually a compendium of a number of articles prepared by two contributing editors of Electronic Design. *Emphasis is placed upon the architectural issues and tradeoffs, the various manufacturer and industry standards being developed, software issues, and applicable protocol design considerations.*

9

The Technology of Implementing
Local Networks

THIS CHAPTER:

will discuss the various technological approaches to local
networking.

will demonstrate the ways packet broadcasting leads to
efficient solutions to the problems of local networking.

Potential users of local networks are undoubtedly confused by the array of new systems, products, and techniques, announced almost daily by dozens of companies, large and small, all claiming to have the definitive solution to the problem of local networking. The development of local networks has been stimulated by the relatively easy entry into the marketplace since the availability of capacity and simple protocols results in low-cost development of the necessary specialized hardware and software. Local network products have evolved from many different types of organizations, including providers of telephone and general purpose communications, and manufacturers of specialized office equipment as well as general purpose computers and data processing equipment. In addition, the large potential market and ease of entry have encouraged the formation of many new companies specializing in local network equipment, design, and installation.

TECHNOLOGICAL CHOICES

The various technological approaches to local networks are distinguished by the network topology and by the transmission medium used to connect the various workstations of the network. The topology, or architecture, concerns the physical arrangement and connectivity of the network's elements. We will consider four basic architectures: the star, mesh, bus, and ring or loop. The four architectures are illustrated generically and schematically in Figures 9-1a through 9-1d;

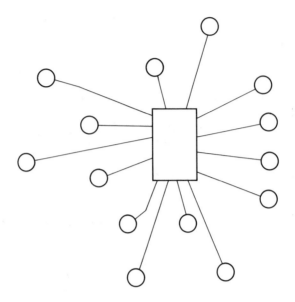

Figure 9-1a.   The star network.

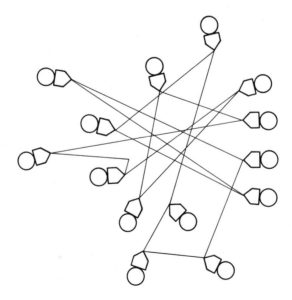

Figure 9-1b.   The mesh network.

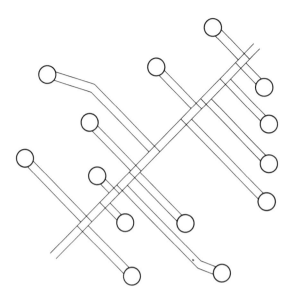

Figure 9-1c.   The bus network.

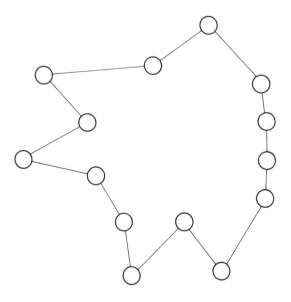

Figure 9-1d.   The ring network.

more detailed configuration diagrams of the various topological structures will be shown as the specific techniques are described in the following sections.

In addition to the four basic architectures, four different types of transmission media are available for local networking. These include inherently shielded cables, such as coaxial cable; nonshielded cable, such as twisted pair wire; fiber-optic cables; and radiated unguided media such as radio, laser, or infrared optics. Table 9-1 indicates practical combinations of network architecture and transmission media.

Superimposed on the matrix of variations are additional technical choices. Coaxial cable systems offer a choice between baseband and broadband systems. The technical distinction between the two techniques will be explained in later sections. For now, we will say simply that, in **baseband** systems, user bits are essentially placed directly on the wire or cable, whereas **broadband** shares technology with cable television systems and requires transmission modulator-demodulators. Such industry heavyweights as Xerox Corporation have chosen the baseband technique, while Wang Corporation prefers broadband.

Ring and loop systems offer a choice of operational protocols, including the carrier sense multiple access, techniques with which we are already familiar, as well as token passing and head-end polling. Very credible organizations, such as IBM and Prime Computer Corporation, have local networking product lines that embody ring and loop systems, with each kind of operational protocol represented.

Though all credible vendors will admit to some shortcomings in their technical approach, all are likely to claim that their approach offers superior advantages offset by only minor disadvantages. Inevitably, the potential user or buyer of a system, or the product development manager trying to decide on the technique to be employed in a new product line, is likely to be overwhelmed by the choices available. In the remaining sections of this chapter, we will attempt to

Table 9-1. Useful Combinations of Local Network Architecture and Transmission Media

	LOCAL NETWORK ARCHITECTURE			
	Star	Mesh	Loop (ring)	Bus
Media				
Coaxial cable	X	X	X	X
Twisted pair wire	X	X	X	X
Fiber optics	X	X		
Radio/optical	X			X

clarify this situation somewhat by presenting a description of the technological choices and their relative advantages and disadvantages.

THE TOPOLOGY AND ARCHITECTURE OF LOCAL NETWORKS

Though the four basic local network architectures all have a number of variations, substructures, and interrelationships to the media in their implementation, our discussion will focus on the basic physical and topological relationships among the elements of each network. In addition, we will show the interaction between the topological design and the resource sharing protocol since in many cases the network structure greatly enhances the applicability of some of the distributed control protocols that were discussed in earlier chapters.

The Star Configuration

Figure 9-2 shows the **star** configuration of local network components.

Centralized Control. The focus of the star structure is the central switch/controller, to which each user device in the network must be connected. Though such configurations are frequently found in data communications or automated office

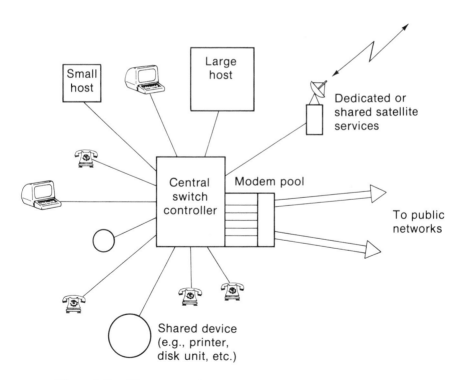

Figure 9-2. The star configuration of local network components.

situations, star networks in local service are most commonly based on modern digital telephone private branch exchange (PBX) switches. With the recent deregulation and growing competition in telephone equipment and services, many innovative modern switching machines have been developed to meet the needs of complex organizations for both basic telephone services and nonvoice digital data services. In such applications, each user device is connected directly to the centralized switch, thus eliminating the problem of capacity allocation or resource sharing, as long as a closed medium such as cable, wire, or fiber optics is used. It is possible to visualize the use of an open medium, such as radio or microwave channels, between the user devices and the central switch; in this case a capacity allocation technique such as time-division multiplexing or carrier sense multiple access, would be required. However, use of such open media is not very likely in the typical local network application, within a single building or complex of buildings.

The central switch in the star topology performs all the processing functions within the local network. Connections between the various local users—which, for nonvoice users, would include any conversions between formats, protocols, speeds, and so forth—are made via the switch. However, this capability actually exists in only a very few commercial products. In most cases the central controller is a transparent device, creating pathways over which the communications information flows between the user end devices.

For voice communication this is what is desired. The majority of the centralized PBX switches use digital operation, but the digitization process occurs in such a way as to allow efficient operation of the circuitry without affecting the quality and characteristics of the speech information. When such devices handle data communications, they generally create transparent connections between the end devices, using the high-speed digital pathways that could also pass voice signals. The bits of the sending user simply pass, essentially unchanged, through the switch, exiting on the path to the destination user.

As a result, such architectures are generally applicable to situations where compatible end instruments need to interchange information. In the local network environment star configurations permit the sharing of certain unique devices that are required to interface the local devices and terminals to larger, broader based networks. Thus, we see in Figure 9-2, between the central local switch and the external networks, a group of shared modems and protocol converters. The centralized switch is generally sufficiently intelligent to be able to determine the kinds of devices that are required to affect certain kinds of interface translations, and to make such equipment available on a shared or pooled basis rather than having to dedicate such equipment to each user on a full-period basis.

Integrating Voice and Data.   The single biggest selling point of centralized star configured local networks is their ability to integrate voice and data services. Let us examine this claim carefully.

While it is true that, by virtue of the digital operation of the centralized switch and the direct connection of each user device to the central point, voice and data signals can indeed pass over the same system, the degree of integration varies greatly among the various implementations of this architecture. In most such systems, the voice signals are digitized into bit streams operating at either 56,000 or 64,000 bits per second, carried to the central switch over several pairs of twisted pair wire for distances of up to several thousand feet. Nonvoice digital signals, at rates up to 56,000 bits per second, could therefore be sent over the same facilities. In some systems, the data signals can operate at the same time as the voice signals; in other systems, however, the data signals have to displace the voice signals. In some systems, a digital data signal, even at as low a rate as 300 bits per second, will displace a full-voice signal channel—in effect using the full 64,000-bits per second capability in the switch to transmit the 300-bit per second information. In other systems, the high-speed voice channels are submultiplexed, so that only a proportionate share of the channel is used to carry lower speed data signals.

Depending on the configuration of the individual ports on the centralized switch, as well as the technology of the local connections between the switch and the end users, some systems can in fact maintain simultaneous voice and data communication. In an ideal system, it would be possible to have two users talking to each other while at the same time sending moderate-speed facsimile or graphical images to support their discussion. In somewhat less sophisticated systems, two users talking to each other could find that they need to send facsimile or graphical images. They could then temporarily suspend or interrupt their conversation and, over the same circuit, transmit the required data. Completion of the data transmission would permit resumption of the conversation. A third possibility is that two users engaged in conversation could exchange some digital data on the same system by establishing a completely separate connection between their two data devices through the centralized switch.

In all three cases, the vendor would be likely to claim integrated voice/data service since the same centralized switching device is used for both types of services. However, the total cost of the different approaches will vary greatly, depending on the costs of the centralized switch, the kinds and quantity of different end-user devices required, and the type of medium used to connect each of the endpoints to the centralized switch.

Data-Only Star Configurations.   There are many applications of centralized star configurations to data devices. Since devices such as shared multiplexors or concentrators are able to perform local switching functions easily, they often form the basis for local data-only star networks. A typical situation is shown in Figure 9-3, where the basic function of the statistical multiplexor is to permit the large number of local terminals to access the limited number of total connections to a remote large-scale computer through the public switched network (PSN). As the

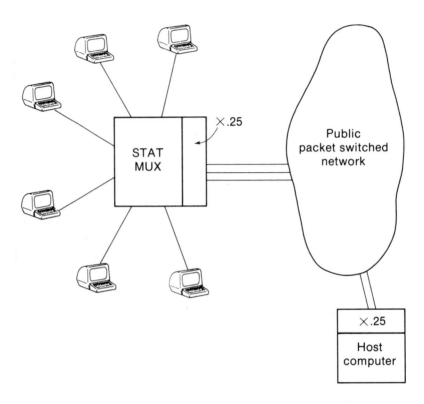

**Figure 9-3. Star configuration data-only network using a
shared multiplexor.**

complexity and capabilities of both the user terminals and the statistical multiplexor have increased, the multiplexor has increasingly been able to function as
the central controller in the local user environment.

As user terminals have taken on more functions than just remote time sharing
to the distant computer, they have needed to exchange information locally with
other users within their own complex. At the same time, it has become efficient
for the centralized device, which was originally just a multiplexor, to perform the
process-oriented functions, such as local switching, and possibly data protocol and
speed and format conversions either within or external to the local communications
system.

Advantages of the Star-Configured Local Network. The major advantages of the
star configuration used for local networks are:

Conceptual Compatibility with Basic Telephone Services. Even in systems
that do not integrate voice and data, the configuration is similar to that of
traditional basic local telephone services. Wiring for telephone service can

be augmented for the data services, or some of the integrated services can even suffice for both the voice and data services.

Independent Operation of Each User End Device.   Since each user has a direct connection to the central switch, each user's operation is independent of the continued operation of any other user's end device. Users do not compete for local capacity, and the failure or improper operation of any particular user's device is not likely to affect any other users.

Presently the Best Vehicle to Integrate Voice and Data Services.   Since most telephone PBX switches are designed for star configurations, and many are capable of handling data services as well, the star configuration is often the only choice for integrated systems. Star-based local networks are often cost justified by the savings and features for voice-based telephone services alone. The incremental cost of combining data services within the same system is very low.

Centralized Control, Administration, Management, and Maintenance.   Since all the network complexity and intelligence is concentrated at the centralized switch, all the network control and management functions can take place through that single device. Because of the centralization of the system information, it is easier to isolate problems and identify them with each user location.

Transparent Operation in Real Time.   The centralized switch creates a fixed path between the two user end devices, which can then communicate with each other over their direct connection. Thus, their interaction is full time, with no processor-imposed delays or reduction in throughput due to competition for network transmission resources. The data comes out of the system in the same way that it entered the network, without concern for delay, ordering, speed, or format alterations.

Centralized Switch Acts as a Focal Point for Shared Facilities and Devices. The centralized switch, embodying a great deal of processing power, acts as a natural focal point for other services that may be required occasionally by the network users. Thus, for example, modem pools can be formed to serve the relatively small number of users who at any time need to access external public network facilities. In addition, the processor capability to do other functions, such as message storage, can be located in the centralized switch or included in a separate processor under the control of the centralized switch.

Disadvantages of the Star-Configured Local Network.   The major disadvantages of the star configuration used for local networks are:

Excessive Resource Requirements Needed to Connect All Users to a Central Point.   Since the star configuration requires a direct connection from each user back to the central point, a very large amount of wire or cable is needed. As a simplified example, we can compare the connection configuration for a single floor of an office building using the star configuration (Figure 9-4a) with a multidrop bus connection (Figure 9-4b). More than 1200 feet of cable is needed for the star connection, whereas only about 400 feet is required for the multidrop bus.

Overconcentration of Resources and Vulnerability to Failure.   The concentration of the network intelligence within a single device through which all communications activity flows leaves the network highly vulnerable to the possible failure or improper operation of the centralized switch. While certain critical elements within the central switch can be made redundant in order to minimize outage possibilities, the system is nonetheless dependent on its proper operation. When the single device is used for both voice and data, it is possible to lose both services at the same time. Since software or data-base problems can cause serious network failures, hardware redundancy does not assure continuous operation.

Limitations on Bandwidth and Data Rate for Each User.   Since each port or connection on the centralized switch has to handle an individual connection, the maximum rate per port is limited. The limitation on most systems is the 64,000- or 56,000-bits per second rate associated with voice-connection data

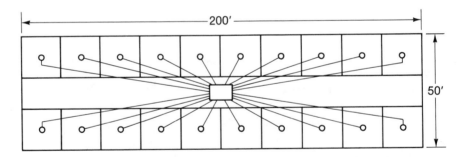

Figure 9-4a.   Star architecture in an office building.

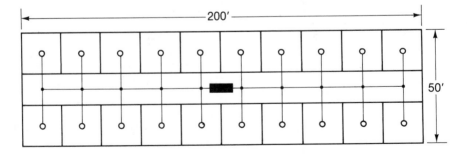

Figure 9-4b.   Bus architecture in an office building.

rates, or the 4 KHz bandwidth associated with the voice channels on inherently analog systems, but in many cases, particularly data-only centralized controllers, the rates may be significantly lower (a maximum rate of 9600 bits per second is not uncommon in data-only systems). The limit is necessary, first of all, to protect the central processing functions from overload from the aggregate input rate of all of the service ports, and secondly, to keep the cost of each port on the central device low.

The Mesh Configuration

Figure 9-5 illustrates the **mesh** configuration of local network components. This approach is a direct transliteration of the concept of connected, long-distance, terrestrial networks to the local network environment. It is particularly applicable to nonhomogeneous systems that begin communicating with each other, on a pair-by-pair basis, and gradually grow into a larger network system. It would also be applicable where the local network spreads over a somewhat larger geographic area than a single building or industrial park. The latter case may call for a combination of different media, such as cable within a building but microwave or optics to locations several miles distant. In such a situation, the various media and

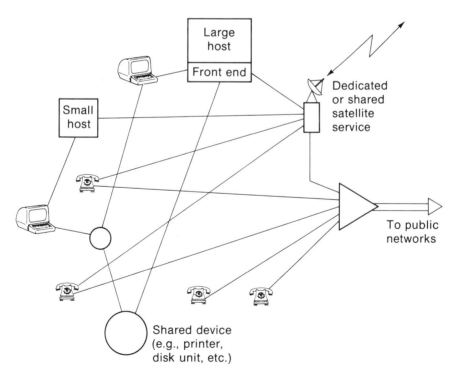

Figure 9-5.   The mesh configuration of local network components.

transmission paths used within the local network are particularly important. Mesh topologies also function well where the local network is an extension of a much larger network; components designed for the long-distance network can be used directly in the local network, possibly with somewhat simpler interfaces or operational software.

The principal strength of the mesh structure is that its resources, primarily transmission capacity, can match the actual network traffic. On the other hand, since not every point in the mesh network is equally accessible from every other point in the network, routing decisions must be made within each node of the mesh in order to insure that the traffic actually reaches the intended destination. Overall, the mesh topology traffic flow mechanisms are the most complex, following the principles of nationwide long-distance networks rather than capitalizing on the high capacity readily available in local networks.

Advantages of the Mesh Configured Local Network.   The major advantages of the mesh configuration in local networks are:

Highly Efficient Use of Resources.   The transmission resources used in the local network, though of low cost, are still a tangible expense. When transmission resources pose a particular problem—for example, where the local network extends over several miles rather than just a few hundred or few thousand feet—the mesh configuration may be the only practical approach. The capacity of the various connections can be matched to the anticipated user needs, resulting in a more economical network design.

Mixed Media Are Readily Accommodated.   When the variation in capacity demands calls for different media, including both closed (e.g., cable) and open (e.g., radio/microwave) media, the mesh provides the most flexible and efficient topology. For example, very short connections between user devices and the nodes of the local network can be low-speed, direct wire connections, while the connections between local network nodes can be high-speed radio or optical links.

Components Designed for Long-Distance Network Can Be Used Directly or in Simpler Versions.   Since the mesh architecture is a microrepresentation of the long-distance network structure, the components designed for such networks can usually be used for local networks without any modifications. In other circumstances, less complex, less costly versions of such devices can be adapted to local networks, while retaining operational features developed for larger nationwide networks for use within the local network. For example, automatic avoidance of failed nodes will be just as valuable in the local network as in long-distance networks.

Intelligence Is Distributed to Nodal Devices Without Burdening Every User Device.   In contrast to the star configuration, the mesh configuration distributes the network intelligence to a relatively large number of nodal devices within the local area. However, individual user devices still see a simple interface to the nodes. The bus and loop architectures, in contrast, require much more intelligence at each user device. Thus, the mesh network is relatively immune to processor failures or anomalies without unduly burdening the individual user end devices.

Disadvantages of the Mesh Configured Local Network.   The major disadvantages of the mesh configuration used for local networks are:

Overall Network Complexity Is Greater Than Other Techniques.   Since the individual nodes of the local network mesh have to deal with various routes, media, and data rates within the network, the nodal devices are inherently more complex and sophisticated than in other architectures. Each node must decide how to handle each message and how to route it to its ultimate destination. Because the various transmission facilities will, in general, have different data rates, the nodal devices have to perform data rate translation, which implies the need for buffering and temporary storage at the nodes.

Actual Traffic in the Mesh Network May Not Match the Initial Design.   Since traffic patterns are likely to change rapidly in a local network (for example, the use of a single high-speed printer among several users can rapidly shift hundreds of kilobits per second of usage to different parts of the local network), the actual traffic flow at any time may be considerably different from that used for the network design. When the flow does not match the facilities well, the result can be reduced throughput, increased delay, and loss of network efficiency.

Failure of a Node Disrupts Service to Many Users.   Though a node failure in the mesh architecture is much less troublesome than a failure in the star configuration, a single node failure will nevertheless disrupt many users. Directly affected are those users connected to the failed node, who cannot send or receive information until the node is back in service. However, also important is the secondary effect of the failure on all users whose traffic would normally pass through the failed node; such traffic would have to be circuitously rerouted, possibly over much lower capacity facilities.

The Bus Configuration

By far the most widely discussed approach to local networks is the **bus** configuration, particularly where advanced office automation systems or local data processing networks are concerned. In terms of installed systems, certainly the star architecture is dominant, since the star is commonly used for most voice-only PBX applications, but the bus structure has gained considerable attention.

As shown in Figure 9-6, the bus architecture simply ties the components of the local network together through a common transmission medium. There is no

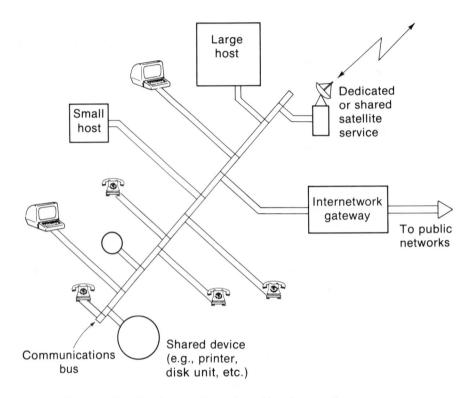

Figure 9-6.   The bus configuration of local network components.

unique control element or central manager of the total resource in the network. The bus structure is a natural extension of the technology used to tie the various processing and storage elements of computer mainframes together. By allowing direct application of the random broadcast techniques discussed earlier, it represents a truly distributed approach to communications.

Bus architectures operate in a variety of ways, using both open media—such as radio broadcast—and closed media—such as wire or coaxial cable—and baseband modulation techniques or broadband techniques.

Baseband Operation.   Baseband operation represents the simplest approach to local networking: The passive, common medium carries the transmitted bits as direct current or bipolar signals. The bit rate on the bus is controlled by the common agreement of all of the user devices, as implemented by the transmitter/receiver interfaces to the bus. Rates of up to 10 megabits per second can be achieved with baseband systems operating over distances of up to about one mile of total system length. The data rate, system length, and transmission medium all interact with each other, and the ultimate limitation on system operation is the ability of the receivers at the distant ends of the bus to distinguish the pulses associated with other user signals, with acceptable error rates, after their transmission along the bus. Since all users are connected directly to the single bus, no routing decisions are made within the local network. The sending station simply transmits its message (or portion of a message, since the maximum length of any transmission is generally limited to a maximum packet length) on the bus. The message flows in both directions away from the transmitting station toward the ends of the bus. Receiving stations, hearing all transmitted messages, select only those addressed to themselves for further processing to the user devices.

Since bus architectures operate without a central controller, and users transmit whenever they have traffic ready to be sent, traffic management has to take place through the contention mechanism. With all of the users operating very close to one another in the local environment, the ratio of maximum propagation delay to packet length is very small. In such a case, as we saw in Chapter 7, the carrier sense multiple access mode of operation leads to achieved throughputs close to full system capacity, with very modest reception delays. In addition, when all users are close together, transmitting stations can immediately detect interference with their transmitted packets. The CSMA protocol is then combined with collision detection, such that a user operates as a 1-persistent packet broadcast station but with the provision that if the transmitting station detects a collision with its newly started transmission, both stations immediately cease transmission and randomly attempt retransmission of the errored packets. Since collided packets are aborted before completion, much less total system capacity is used than with the basic ALOHA channel, or even with full CSMA, where packets run to completion even when collisions occur.

The basic operational protocol for bus-type local network architectures is thus known as carrier sense multiple access with collision detection (CSMA/CD). Since very little capacity is wasted on undeliverable packets, network efficiency is very high, as shown in Figure 9-7. When the network traffic is light, collisions occur very infrequently and those that do are quickly aborted, so useful throughput is essentially equal to total traffic (offered load). When the channel gets heavily loaded, collisions are much more frequent, and persist for times approximately equal to the average propagation delay along the bus, resulting in lost capability. Short packets (500 bits per packet) result in overall efficiencies of about 83%, while longer packets (approximately 4000 bits per packet) can achieve overall efficiencies of about 96%. Longer packets result in higher efficiency since more useful traffic is sent once a transmitter successfully captures the channel, assuming that all other users refrain from transmitting while the channel is in use. However, longer packets naturally lead to much longer user-perceived delays when the channel is heavily loaded.

The major criticism of the CSMA/CD bus architecture is its inability to assure successful delivery of any user's information when the channel is heavily loaded. Under a heavy load, users see long delays because the channel is sensed as busy most of the time, and collisions frequently occur when the channel becomes idle. Also very troublesome is the uncertainty of getting even a short packet's worth of data through the heavily loaded channel.

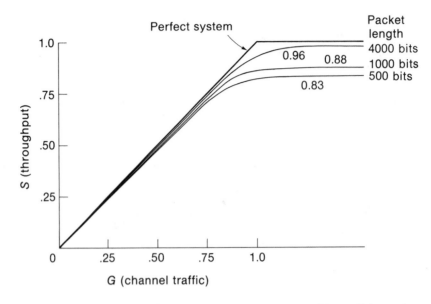

Figure 9-7. Performance of a CSMA/CD network (Xerox Ethernet) under heavy load. (Source: Shoch and Hupp, 1980.)

Broadband Operation. A much more sophisticated approach to the bus architecture is the broadband technique, which combines the concepts of frequency-division multiplexing (FDM) with the carrier sense multiple access, or other forms of time-division multiplexing (TDM) that are applicable to the baseband technique. The broadband technique not only permits much higher capacity on the CSMA subsystem, it permits the combination of local data communications with many other types of services, including very high bandwidth services such as cable television. In other words, broadband techniques permit the establishment of local data networks on facilities that may already exist for other functions.

Broadband techniques employ somewhat more expensive cable than is generally used in baseband systems. More significantly, broadband interface transmitter/receiver devices are far more complex, and cost anywhere from three to ten times as much as the equivalent baseband devices. Broadband systems use the entire radio spectrum bandwidth available between roughly 10 megahertz and 500 megahertz, where the digital signals are frequency or phase modulated on carrier frequencies that divide the usable bandwidth into a number of different operational subsystems.

Just like different channels on a cable television system, local networks and subnetworks operating within the common cable can be assigned to different frequency bands. Narrow bands, of possibly 50 KHz to 250 KHz per band, can be used to support data rates of 9600 to 50,000 bits per second. Wide bands can be used to create channels of 10 to 30 megabits (Mb) per second using bandwidths as great as 35 MHz. Each of these channels can operate as a separate CSMA/CD channel, with the option of combining user access to any or all of the channels—thereby creating systems of huge capacity.

A single cable-based local network could thus be used to establish a large number of small independent systems, using fixed frequency modems with each set of modems assigned to a preassigned operating frequency. For example, in a large office building or industrial park, a single cable system could serve the needs of different companies, each being assigned to its own operational frequency band. Conversely, a very large system, with almost unlimited total capacity, can be created for a single organization, using tunable modems, where modems can be remotely tuned to any of many different operational frequencies on the cable, and possibly the additional complexity of a centralized bus controller and frequency manager. In such a large system, two users who desire to communicate would first interrogate the central controller to agree upon a commonly accessible frequency band. Both users would then program their tunable modems to the same frequency band, where they would communicate using the CSMA/CD protocols. A schematic representation of such an approach is presented in Figure 9-8.

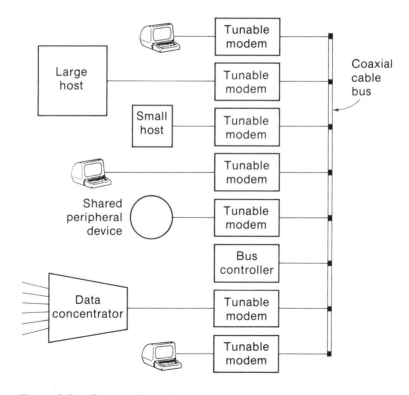

Figure 9-8.   Bus configuration for a single large system using tunable modems and a centralized controller.

Broadband and baseband techniques for implementing local networks are so different that they are not even directly comparable. Each approach is well supported in the marketplace, with the baseband technique used as the basis for Xerox's Ethernet and the broadband technique used by Wang's Wangnet. Many other manufacturers have developed product lines that use one or the other of these technologies, but at present, because of Xerox's marketing, the baseband techniques are more widely used.

Ultimately, the user's decision between baseband and broadband techniques must be based on the total network complexity and capacity required on the local network. Baseband approaches are far simpler to implement and less costly to install, but they have a limited total capacity. Broadband systems are much more complex and expensive, but they can meet a much larger total demand as well as a much wider range of potential applications.

Advantages of the Bus Configured Local Network. The major advantages of the bus configuration used for local networks, in either baseband or broadband techniques are:

Bus Configurations Use Minimum Amounts of Cabling. By use of a single cable, tapped wherever necessary to attach users, a minimum amount of cabling is required within the local network. If necessary, a local network can be extended beyond the normal range of a single cable, by multiple segments of cable connected together and using digital amplifiers or repeaters.

Passive Operation and Distributed Control Insure High Reliability. Since the common bus operates passively, acting only as a medium for the users' transmitter/receiver devices, the medium itself is highly reliable. The distributed control means that failure of any one user device will not generally affect the traffic flow of other users. Of course, it is still necessary for the system to protect itself from the possibility that a device could fail in a constant transmitting state, excluding any useful traffic from the system.

Users Can Be Readily Added, Removed, or Reconfigured in the System. The network structure or operation does not depend on any one user for connectivity or routing of information. As long as the user identification or address is unique, it can tap into the network at any point or at any time, or move to some other point in the network without any rearrangement of facilities.

Bus Architecture Is Ideally Suited to the CSMA/CD Protocol. The passive, low-loss medium provided by the transmission bus is ideal for the carrier sense multiple access algorithm and protocol, with or without collision detection. This protocol achieves high throughput, especially when the ratio of propagation delay to packet length is small (which is generally the case for local networks), and the average delay is acceptable, even in fairly heavily loaded systems.

Much of the Hardware Is Fully Developed and Mass Produced. Bus architectures employ devices, facilities, and components that are widely used in television distribution systems and other cable-based systems (such as surveillance, telemetry/monitoring, etc.) and are thus readily available and relatively inexpensive. Open development of the techniques and free and easy licensing of many of the patented devices and interfaces have stimulated development and competition.

A Single System Can Serve Many Diverse Needs. Because of the large capacity available on the bus, many different functions can be integrated on a single bus system. Particularly in the case of broadband systems, local data networks may take only a very small percentage of the total system capacity.

Disadvantages of the Bus Configured Local Network. The major disadvantages of the bus configuration used for local networks, in either baseband or broadband techniques are:

Geographic Coverage of System Is Limited. Proper operation of the bus system and the distributed interface protocols limit the overall coverage of the network to a mile or so at most.

Distributed Protocols Can't Insure Traffic Delivery. Since the network operation is not managed by any central controller, it is possible, especially under very heavy loading, for delivery delays to become excessive, or users' traffic not to get through the system successfully. Because each user interface needs to operate similarly to other user devices, it is difficult to allocate more capacity to some users than to others, or to give more important users higher priority in the system.

Certain Types of Failures Could Easily Disable the Entire System. The CSMA/CD protocols are very sensitive to problems caused by user devices failing in a transmit state or not reacting promptly to a transmission collision. Interface devices have to be carefully designed to "fail-safe" and disconnect themselves from the bus if their operation is impaired.

The Loop, or Ring, Configuration

A number of approaches connect a local network in a closed, circular pattern. One approach is basically the same as the bus architecture, with the two ends of the bus closed in a loop; the performance of this structure is similar to the bus. The **loop,** or **ring,** configuration preserves the advantages of the bus architecture, with the additional enhancement that, if the bus should be broken or damaged at any point, a path still connects all the users. This approach is illustrated schematically in Figure 9-9.

A second approach to the loop architecture that has been used in local network environments is based on the polling chain, which was discussed in

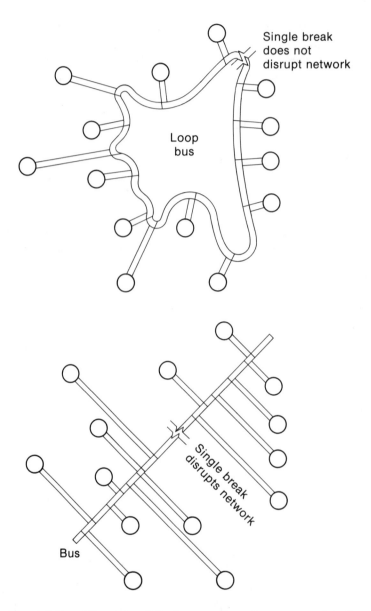

**Figure 9-9.   Advantage of the loop configuration compared to the
straight bus configuration.**

Chapter 3 (see Figure 3-2). In this case, the ends of the polling chain are tied
together in a loop to provide greater assurance of connectivity in case of physical
damage to the connecting wires or cable. This form of the loop usually operates
under a central controller or master polling station. Such loops generally operate

at relatively low speeds since connection may often be over telephone-quality circuits, thus limiting the basic line data rate to 9600 bits per second or less. In entirely local networks, rates are often up to 50,000 bits per second. The "deformation" of a polling chain into a polled loop is shown in Figure 9-10.

The Synchronous Loop, or Ring.   A unique circular local network structure is the **synchronous loop.** In this case, the nodes of the network are not simply bridged onto the conducting medium, they are inserted in tandem with the conducting medium, creating a closed chain of sequential hold-and-forward packet switches. This ring structure can be likened to a closed loop railroad, with frequent loading platforms spaced throughout the system. Box cars move continuously along the track, stopping momentarily at each of the loading platforms. At each platform, freight may be loaded on or removed from the boxcar under certain conditions. Freight may be unloaded only when it is addressed to that particular station. Freight may

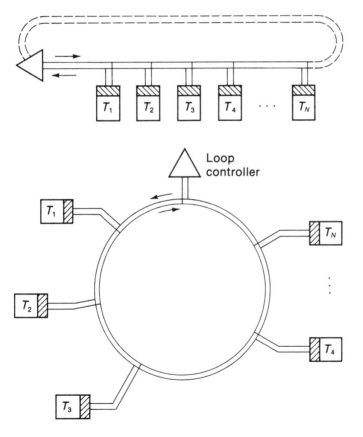

**Figure 9-10.   Deformation of a polling chain into a polled
loop configuration.**

be loaded onto a boxcar only when the conductor at that station is holding a validated ticket, or **token,** authorizing that station to use system capacity at the present time. Otherwise, the box car simply moves on without loading or unloading.

In the same way, the data communications local network in the ring configuration simply passes packets of data, node by node, around the circular connection, as shown in Figure 9-11. The key to the advantages of the ring configuration is the existence of a transmit token (often called a write token), which permits new data to be entered into the network. The token concept prevents stations being locked out of the system by competition with other stations under heavy load, which can occur with the CSMA or CSMA/CD algorithms on the local network bus architectures. Since a station can transmit only when it has the token, the algorithm requires that the stations pass the token around the ring on an equal basis, so each

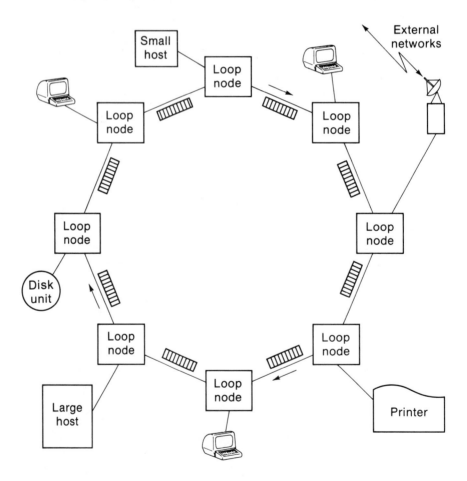

Figure 9-11.   Ring configuration of a local network.

station, in turn, has the opportunity to enter data when an unfilled slot comes by. In addition, loading of the system is quite constant. The average delay is based simply on the time it takes the average packet to travel halfway around the ring. Of course, if the whole system is heavily loaded, it may take a while before the token comes around to any given user since the token is held longer by a user who is transmitting a packet than by one who has nothing to transmit when he receives the token. In a lightly loaded system, where most of the users are idle, the token moves around the ring rapidly.

Advantages of the Ring Configured Local Network.   As a result of the tight control of the network operation and capacity allocation, the ring architecture has the following advantages for local networks:

> *Network Interfaces Are Quite Simple.*   The protocols for entering and removing data from the ring are quite simple, requiring considerably less complexity than the carrier sense techniques.

> *Control Is Distributed and Capacity Is Equally Allocated.*   The distributed control algorithm with token passing insures that operation is not dependent on any single node. The potential for using system capacity is equally distributed among all the nodes, regardless of their position in the system.

> *Performance Remains Constant and Stable Even Under Heavy Loading.*   Since network operation is quite deterministic, with the entry rate of new data controlled by the movement of the write token, the network cannot become congested with undeliverable traffic or packet collisions.

Disadvantages of the Ring Configured Local Network.   The synchronous data loop, or ring architecture, used in local networks has some significant disadvantages, which can be summarized thus:

> *Each Node Is an Active Element in Maintaining Proper Operation.*   The loop consists of many segments tied together by nodal processors. If any point in the loop is broken, either at a processor because of an equipment failure, or in the circuitry connecting two nodes, operation of the system halts. This problem is so critical that most systems use two concentric loops, as shown in Figure 9-12, with data flowing in opposite directions around the two loops. If the line or any processor fails, the other processors are programmed to bridge the two loops together before the break to keep the loop operating.

> *Average Delays Are Long, Even Under Light Load.*   Because, on the average, every message has to pass through half the nodes in the network, delays under light loading are much longer than in other local network architectures. Users are inhibited from transmitting until they find available capacity slots at the same time they hold the transmit token.

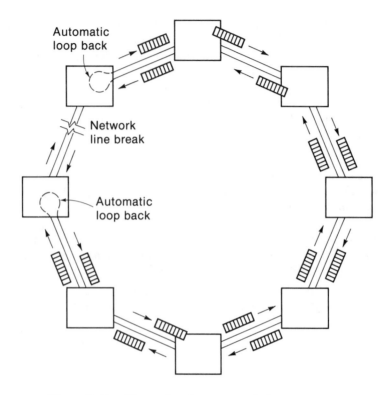

**Figure 9-12.   Network using concentric loops to assure
system operation.**

THE LOCAL NETWORK ARCHITECTURE—ONLY PART
OF THE SOLUTION

The many different forms of high-capacity, high-reliability communications facilities only partly solve the problem of providing communications for a large number of local users. Data bits emanating from a large number of machines can be sent to other machines at rates of many millions of bits per second if necessary. However, the transport mechanism that permits such exchange of data does nothing to insure that those bits will be understood when they reach their destination. This situation has always been accepted without much question in voice communications, where the designer was satisfied to place a reasonably high-quality communication circuit between two users. The fact that a speaker of Japanese could not be understood by a user fluent only in French was seen not as a shortcoming of the system but only as a problem of the users.

In an ADP system, by contrast, the ability of any user to access the information is limited much more by the human–machine interactions than by any person-to-person processes. Consequently, in the design of data processing equipment, both hardware and software engineers as well as applications engineers continually strive for a higher level of commonality. Standardization enhances the potential market of any given product and at the same time increases the utility of that product to any user. Local network architectures provide an efficient data flow mechanism to meet, in a limited geographic area, the demands of the lower three levels of the seven-level ADP/data communications protocol hierarchy.

The International Standards Organization Protocol Hierarchy

In order to facilitate the exchange of information between diverse systems, a layered approach to structuring the networking protocols has evolved. Principally through the efforts of the International Standards Organization (**ISO**), a seven-level standardized definition of communications protocol functions has been developed, as depicted in Figure 9-13. The concept behind standardization is that each level of the hierarchy can be designed independently of the other layers, with the possible exception of the layers immediately above and below any given layer. Levels 1, 2, and 3 involve fundamentally communications functions, whereas levels 4 through 7 involve fundamentally data processing functions. At any given

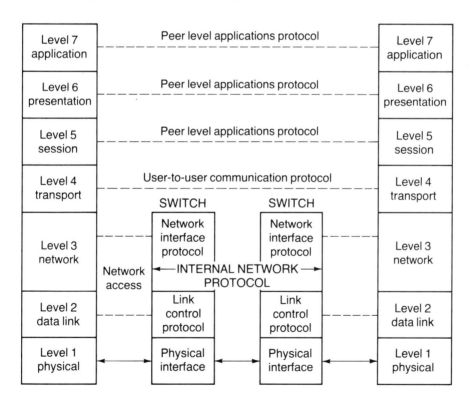

Figure 9-13. International Standards Organization (ISO) protocol hierarchy.

level, two systems sharing a common design and function set would be able to communicate and exchange information at that level, regardless of the remainder of the system design.

The lowest level of the ISO protocol hierarchy is the physical level, where existing standards for communications hardware interface are applied. Hardware interface refers to the pin connections, electrical voltage levels, and signal formats. The second level, known as the link level, refers to the data link between the user and the network. This level defines the data format, error control and recovery procedures, data transparency, and implementation of certain command sequences. For nonswitched networks, or the interface of simple terminals with computers through point-to-point services, generally only levels 1 and 2 are required. Networks designed by a single manufacturer around a single product line generally operate with a combination of level 1 and level 2 protocols.

Level 3, the network level, defines most of the protocol-driven functions of the user interface to a network. It is at this level that the flow-control procedures are employed, switched services are initiated through data call establishment, and messages are decomposed into packets. In local networks, particularly those employing CSMA/CD, the logic associated with packet broadcasting, collision sensing, and retransmission scheduling takes place at this level.

Level 4, the transport level, is required to assure the end-to-end flow of complete messages. If the network requires that messages be broken into segments or packets at the interface to the network, the transport protocol layer assures that the message segmentation takes place and that the message is properly delivered. In local networks, where operation of the network follows primarily a datagram mode, with the possibility of losing, duplicating, or misordering packets, level 4 would assure that messages are properly reassembled from their component packets.

Level 5, known as the session control, controls the interaction of the user software exchanging data at each end of the network. Session control includes such things as network log-on, user authentication, and the allocation of ADP resources within the user equipment.

The presentation layer, level 6, controls the display formats, data code conversion, and the control of information to and from peripheral storage devices.

Level 7, the user process or user application level, deals directly with the software application programs that interact through the network.

Although at levels 5, 6, and 7, the protocols (often known as peer level applications protocols) are defined from a functional viewpoint, implementation of standard software that can operate at these levels has been slow. The software at all levels tends to be both equipment and application dependent. However, the layered approach to protocol development achieves a degree of isolation and modularity between the various layers, so that changes in one layer can be made without changes in any of the other levels.

Because of the isolation among the levels, as well as the high data rates and large capacity available in the local networks through the operation of the level 1 through 3 protocols, greater operational commonality at the higher levels is being achieved. This results from the ability of the user devices to pass a large amount of overhead and protocol information efficiently among the network elements. Xerox has developed widely applicable specifications for all levels of the Ethernet through level 6, including the ability to recognize manufacturer-unique higher level protocols from manufacturers other than Xerox. Manufacturer-unique protocols could use the network as a transmission medium to interface among themselves, while the publicly available Ethernet protocols could exchange information independent of manufacturer and product line. These goals are more easily attainable in local networks than in broadly distributed networks because the high capacity and low cost of local transmission allow for more elaborate control mechanisms as well as greater redundancy in information content. Consequently, local networks will become more important as a mechanism by which various functions can use common equipment to share common data bases, storage devices, and software. In addition, with a small number of processors to perform the protocol conversion to capacity-limited media, the local network has increased advantage in performing the concentration and interface function between the local and the long-distance network environments.

SUMMARY

1. The selection of an implementation technique for local networks is made particularly difficult by the claims and counterclaims of the many system providers.

2. Four different topological arrangements—the star, the mesh, the bus, and the ring—combined with a variety of transmission media and a number of different operational protocols give a wide range of local network configurations.

3. The star configuration evolves readily from the typical telephone private branch exchange (PBX), with modern, digital centralized switches permitting a large degree of data and voice integration.

4. The mesh architecture can accommodate a wide range of media and facilities within a single system, and can evolve over time using many of the facilities and techniques of longer distance, large-scale networks.

5. The bus architecture, either in the baseband or broadband form, provides for very efficient connection of network components, basically passive operation, and highly flexible configuration management.

6. The ring architecture uses a closed configuration with very high reliability, simple user interfaces, and constant, stable performance.

7. Commercially available local networks provide well-established implementation of the lower three levels of the ISO protocol hierarchy for interconnected ADP systems. They will contribute materially to the evolution of broadly applicable ADP and office automation/information-processing systems.

SUGGESTED READING

SHOCH, JOHN F., and HUPP, JON A. "Measured Performance of an Ethernet Local Network." *IEEE Symposium on Local Area Communications Networks*, Boston, May 1979. (Available from Xerox Palo Alto Research Center, 3333 Coyote Hill Road, Palo Alto, California 94304.)

This report compares actual Ethernet performance to the predicted characteristics. Typical user characteristics are also shown in terms of how the network performance affects their utilization of the network.

THURBER, KENNETH J., and FREEMAN, HARVEY A. "The Many Faces of Local Networking." *Data Communications*, vol. 10, no. 12 (December 1981), pp. 62–70.

A broad introduction to local networks, this article emphasizes bus architecture using broadcast multiple access techniques. It goes into some detail on the implementation used in the Xerox Ethernet approach to local networks, and the ability to interface a number of different bus and nonbus based networks. The article is enhanced by the inclusion of a comprehensive list of manufacturers and suppliers for local network products.

10

Security, Privacy, and Protection in Distributed Communications Systems

---

THIS CHAPTER:

will explain the threats to unprotected information carried by distributed communications systems.

will describe the widely accepted Data Encryption System.

will discuss both administrative and technical protection of information in distributed communications systems.

Wiretapping has always been a widely recognized threat to the privacy and security of information transmitted through communications networks, but the risk of interception is magnified manifoldly in shared communications resources and distributed networks. The very technology that provides the many desirable features of such networks can be used to conduct sophisticated "attacks" on the system. Fortunately, the technology can also protect those facilities, and at the same time make user information even safer from intentional and unintentional loss or disclosure than it would be in less economical, nondistributed communications networks.

The required controls add significant expense to network implementation as well as additional complexity to network design and operation. However, those costs must be measured against the cost of possible compromise of the information or the potential economic losses from undetected fraud or theft of property or information.

THREATS TO COMMUNICATIONS SYSTEMS

Information technology, particularly as applied to modern office automation systems, in large measure replaces paper with electronic storage, manipulation, and distribution of information. As a result, the information becomes vulnerable to three types of threats: interception, insertion, and disruption. We will consider each in turn.

Unauthorized Interception of Information

Two users of communications facilities, will normally assume that they are the sole listeners to the information being transmitted. And in fact, the privacy of the information is fairly high because of the practical difficulties of intercepting communications and associating the information with one particular user. In the switched telephone network, for example, it is relatively easy to intercept literally thousands of conversations by receiving the signals transmitted from hundreds of microwave towers around the country. However, it is highly unfeasible to selectively intercept the information of one particular user, since, at any time, that user's communications could be on any one of thousands of channels carried on hundreds of possible routes. Physically "wiretapping" the actual wires or cable close to the user's end instrument requires access to the user's premises, which is often very difficult. Consequently, as shown in Figure 10-1, conventional communications systems are relatively safe from interception.

In satellite-based or other distributed communications, the situation is quite different. As we have seen, in order for the systems to operate efficiently, each user's

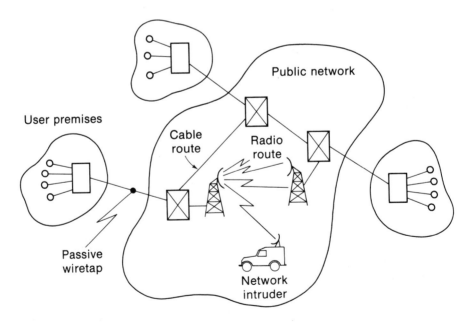

**Figure 10-1.   Vulnerability of a conventional communications
network to interception.**

information must be clearly marked with both source and destination information. Consequently, in national long-distance networks, information can be received almost anywhere in the network and easily associated with the proper source and destination users. Even in local networks, the bursty operation typical of most systems makes the information readily identifiable. In addition, because the information is transmitted digitally, computers can easily search and analyze very large volumes of data. As a result, information in distributed data networks is vulnerable to interception.

Intercepted information could be used to obtain economic, market, or technical data from a competitive organization; or economic, legal, or regulatory information from various government agencies. Interception may entail access to private and personal information about individuals or infringement on citizens' rights by government. In any case, information interception, long the mainstay of industrial espionage and political intelligence gathering, would be made significantly easier by the applications of distributed networks unless specific safeguards were placed within the network.

Unauthorized Insertion of Information

The ease with which users can be added, deleted, or moved within distributed networks significantly contributes to the problem of unauthorized insertion, modification, or deletion of information. This vulnerability, often known as **spoofing,** has resulted in some of the most spectacular computer and white-collar crimes ever reported, and undoubtedly many millions of dollars of losses are still unknown or unreported. The widespread use of distributed data communications networks for electronic funds transfer, credit card validation and accounting, computerized teller machines, and order entry facilities makes it relatively easy to insert unauthorized information into such a system and thus improperly transfer money or property.

For example, as shown in Figure 10-2, a wiretapper could insert packets into a distributed electronic banking system by monitoring the network for a while and then mimicking the format of the legitimate packets. Using such packets, he is then able to improperly transfer funds or resources between properly established accounts within the institutions on the network.

Intentional or Unintentional Disruption of Information

Perhaps the most insidious threat is the loss or disruption of information through either intentional or unintentional abuse or misuse of the distributed system. Even properly authorized users could, through misuse or malfunction, render the system useless or significantly disrupt the information stored in it. The most graphic example of this situation is in the multiple access systems, such as the CSMA/CD, where the failure of a system component in a constantly transmitting state would render the entire system useless until such a user device could be repaired or at least isolated from the network. Users of such facilities could

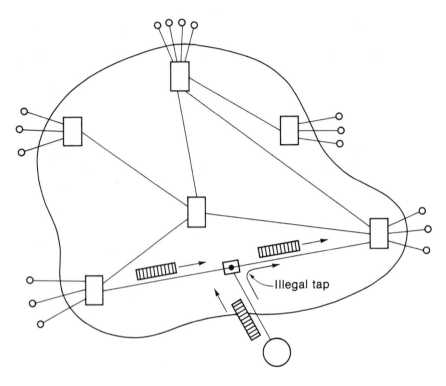

Figure 10-2. Insertion of packet message into a distributed network.

conceivably modify their systems in order to capture an unfair share of overall system capacity by, for example, shortening their terminal retransmission delay or increasing their transmitter power. Unauthorized users could attempt to jam the network to disrupt legitimate use, or damage or destroy the data bases and end-user information accessible through the distributed network.

Internal Threats to Information Privacy and Security

In addition to the external threats to security we have described, a wide range of threats result from the potentially large number of authorized users of distributed communications networks. Through software errors or hardware failures, information could be altered, misfiled, or lost on entry into or exit from the network. Users could use the network to interrogate or probe files of other users, or to access information to which they were not entitled.

Internal threats may also arise from intentionally placed software features that allow systems programmers to access the network or modify its operation without full visibility to or authorization from the network management. Such threats, often referred to as **Trojan Horse** threats, are particularly plausible in an environment of high job mobility, where a programmer may soon be working for a competitor of his current employer. Trojan Horse threats are extremely hard to detect and to protect against, as was demonstrated by the long-term theft of many millions of dollars of pari-mutuel winnings in Florida by several systems programmers who knew the design of the betting system.

Whereas external threats to information security can be met with a number of technical controls, which effectively limit access to information in the system to authorized persons, protection from internal threats must generally come from nontechnical controls. Personnel security must be carefully maintained, software checks and cross-checks must be used to detect intentional or inadvertent changes to the expected operation procedures, and departure of key personnel must be followed by changes in access controls.

TECHNICAL APPROACHES TO NETWORK SECURITY

The "Combination Lock" Principle

Like the vault in a bank, information and communications security has been based on the concept of a secret combination. The sender of the information locks the transmitted information with the secret combination, or key, and the receiver, the only other person with knowledge of the key, can unlock the information on receipt. As in the case of a vault, it is theoretically possible to hit on the right combination by trial and error. However, if the combination has a sufficiently large number of digits, the number of possible combinations makes it practically unfeasible to try to open the vault this way.

For example, in a simple combination lock employing three numbers, each with two digits, a total of 1 million combinations exist, ranging from 00-00-00 to 99-99-99. Assuming some one knew the correct directions to turn the dial, and tried one combination every 6 seconds, it would take more than 69 days, working 24 hours a day, to try all the possible combinations. If the lock were extended to five numbers, each of two digits, it would require more than 1890 years to try all the combinations, even if ten tries per minute could still be achieved.

Advantages to Authorized Users

Applying the same principles to communications links provides much the same advantages to the sender, as long as the key is kept secret. Figure 10-3 shows the elements of a typical **encrypted communications link.** For simplicity, let us assume that the messages to be sent from terminal A to terminal B are always in the form of a series of numbers. To protect these messages, A and B agree to use an **encryption** key, based on a mutually agreed-on six-digit number, such that each message from A is multiplied by an encoding key, K. Thus, instead of transmitting a given message (plain text), say M_1, user A actually transmits the encrypted message $M_1K = M_t$, known as the ciphertext. At the receiving end, the received

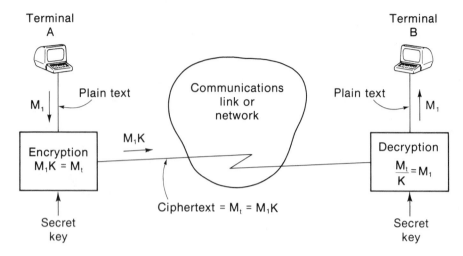

Figure 10-3.   Elements of an encrypted communications link.

message, M_t, is divided by the agreed-on key, K, in order to **decrypt** the ciphertext, such that $M_t/K = M_1$, the originally transmitted message. Any intruder listening to the channel, not knowing the key, K, could decode the message only by trying every possible key value. In addition, the intruder would also need some sample of both clear and encoded text to match to each other, in order to know when the correct key was found.

The economics of the encoding process greatly favor the authorized users of the communications link. Assume, for example, that the communications link users are willing to spend 10¢ to encode each message. In 1960 an investment of 10¢ would permit the processing of approximately 5000 bits, or, in effect, an encyphering process permitting the manipulation of this many bits. Without the secret key, the decoding process would require at least $(5000)^2$, or 25,000,000, bit manipulations, entailing a processing cost of about $500. By 1980, the trends in data processing costs made it possible to process about 20,000,000 bits for 10¢. The required decoding by an intruder without the secret key would require $(20,000,000)^2$ operations, or a total of 4×10^{14} bit operations, for an estimated processing cost of more than $2,000,000. If the decoding processor could manipulate 100 million bits per second, it would take the processor more than 46 days, operating 24 hours a day, to break the code of a single message.

From all points of view, it is relatively easy to encrypt, and much harder and more expensive to decrypt, unless the receiver has the proper secret key.

The Encryption Process

The encryption process takes place as a mathematical manipulation of the sender's message, with an inverse process taking place at the receiving end. With binary communications, the mathematical process can be simply binary addition of a randomly chosen sequence with a similarly long portion of the message. If the randomly chosen key is at least as long as the message, and used only once, then the message is unconditionally secure and cannot be broken regardless of processing power. However, this is not a very practical way to perform the encoding. As we have seen, a finite-length key can be sufficiently long to insure conditional security; that is, determining the proper key by an exhaustive search is impossible in a practical sense. With a finite-length key and a relatively simple mathematical operation at the sending and receiving end, a high degree of protection can be achieved. The encryption can be applied on either a link-by-link basis or an end-to-end basis.

In end-to-end encryption (Figure 10-4) two users of a distributed network each apply encryption devices at their terminal locations, agreeing on the secret key to be used. End-to-end encryption provides the greatest degree of protection since the information is fully encoded all through the network, except at the users' own end terminals. However, end-to-end encryption is the most difficult to implement. In order to insure proper operation of the decoding process at the receiving end, the decryption mechanism must be synchronized with the encryption mechanism. There are a number of methods of doing this, but each is rendered less effective by the presence of variable delays. In addition, if the network uses switching, there must be a method for transmitting the address and signaling

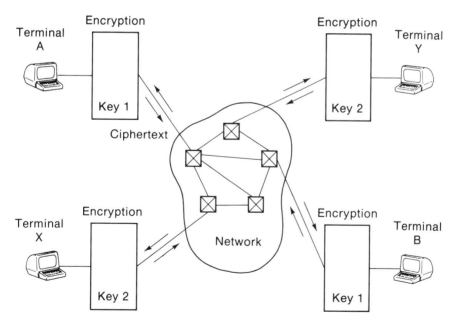

**Figure 10-4.   End-to-end encryption. Terminals A and B exchange
protected information using key 1; terminals X and Y
exchange protected information using key 2.**

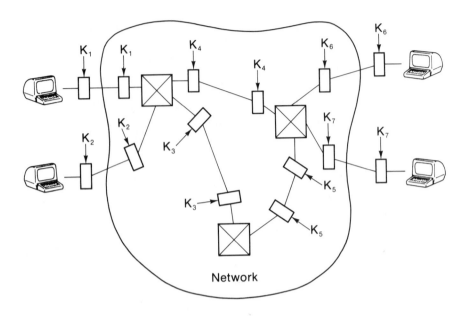

Figure 10-5. Link encryption.

information in the clear, so that the overhead information can be understood by the network switches and control facilities.

Link encryption (Figure 10-5) is applied on each line segment in the distributed network, such that all information flowing on the communications line is unintelligible to an unauthorized intruder. Even the overhead, address, and control information are protected, offering traffic flow security as well as information security. (Traffic flow security means that an observer does not even know if and when information is actually being transmitted on the communication line.) The major disadvantage of link encryption is that the information has to be decoded at each of the nodal elements of the network, so the information would be intelligible to an intruder who has penetrated the switching or processing elements of the network. Link encryption also requires more encryption devices in a typical network since the lines between the switches and nodes, as well as the lines between the users and the network, have to be individually encrypted. However, in truly distributed networks and networks based on broadcasting techniques, the distinction between link and end-to-end encryption are minimized since, for the most part, the traffic moves directly between the source and destination users.

Key Management

The security of a distributed network and its users' information is only as good as the secrecy of the keys used in the encryption and decryption process. In link-encrypted networks, the key distribution function can be accomplished with a minimum of technology since only one pair of keys is needed for each link in the network. Though the compromise of a given key will potentially compromise all traffic on any particular link in the network, this is only a small fraction of the total information in a multiple-link, multiple-route network.

End-to-end encryption presents a much greater challenge as, in effect, a separate key must be generated for each pair of users, requiring each end user to keep a secret catalog of keys for each other network user. If a different key were not used for each user pair, compromise of the key would compromise all the traffic throughout the entire network. A possible alternative for end-to-end encryption is to establish communities of interest that share a common key. A given user might be associated with ten different communities of interest, and would have to store the keys associated with each different community. Messages destined for other users within a given community would be encrypted using the common secret key assigned to that community. This significantly reduces the total number of keys required, while still minimizing the amount of traffic exposed in case of the compromise of any single key.

A sophisticated approach to end-to-end encryption, which will become more practical with the further reduction of processing costs, uses a centralized key management facility (Figure 10-6). Users are each assigned long-term individual keys, which they use to communicate with the key management facility. The key

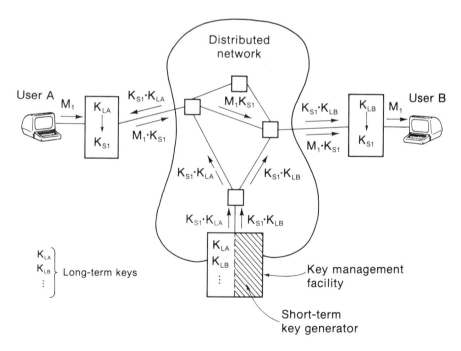

Figure 10-6. End-to-end encryption using a key management facility.

management facility, in reality a specialized user computer attached to the network, stores the long-term key for each user in the network and can communicate on an end-to-end secure basis with each user. In order to transmit data through the network, a given user would first transmit a request to the key management facility to establish a new, encrypted connection through the network. The key management facility would randomly generate a new short-term key and transmit it to both the requesting user and the destination user. The destination user, upon accepting the new network connection from the originating user, would receive a copy of the short-term key, generated just for that connection. Since the key management facility can communicate with all users in the network under end-to-end protection, the distribution of the short-term keys from this facility is clearly protected from interception, while it allows each user an individually protected connection with any other user in the network.

Public Key Systems

Another approach to key management permits widespread public distribution of a nonsecret encryption key, which greatly simplifies key management and control. All users employ the same encoding algorithm, and originating messages are encoded in the public key associated with the destination user. Thus, for example, the network user directory would include, in addition to the destination address of each user, each user's unique—but public—encryption key. Any messages sent to that destination would be encoded using that same public key.

The viability of public key systems is based on the theory of **trapdoor,** or noninvertible, **functions.** The concept uses the fact that, although certain functions are relatively easy to compute in the forward direction, they are very difficult to compute in the reverse direction. For example, it takes just a few seconds to compute $(123)^3$ and find that it is 1,860,867; it takes considerably longer to determine that the cube root of 1,860,867 is 123. By employing relatively large numbers, such as the product of two 50-digit prime numbers, it is possible to choose an exponent for the encoding process that makes it mathematically unfeasible to invert the encoded message unless the receiver also knows the two prime numbers that were used in generating the exponents used in the encoding. And if the number is sufficiently long, it is also mathematically unfeasible to factor it back into its two prime components. For example, if a 200-bit long number is generated by the product of two 100-bit long prime numbers, it would take a computer with the ability to process 1,000,000 instructions per second more than three days, operating full time, to factor the number back into its prime components. Without those prime components, it is unfeasible to compute the inverse of the encoding exponent. If the numbers were lengthened to 150 bit-long prime numbers, it would take more than three years for the factoring process, and with 250-bit prime numbers, more than 1 million years.

Public key systems will undoubtedly become widespread over the next decade, as more and more public financial transactions proceed over communications facilities. The many forms of electronic funds transfer and credit validation and authorization systems mandate a high level of security together with a high degree of individual protection for a large number of users. The time will probably come where individuals will be assigned a public encryption key, just like a social security number, to be used in their own transactions.

The Data Encryption Standard (DES)

At the present time, and for the near future, the most practical and cost-effective technical approach to communications security follows the **Data Encryption Standard** developed by the U.S. Bureau of Standards and endorsed by the U.S. government. Developed in the early 1970s by IBM under contract to the National Bureau of Standards, the DES was adopted in July 1977. It provides a standard method of employing link encryption so that any device that follows the specified algorithm can decode the information encoded by any other device operating under the standard, as long as the proper key is known. The standard balances complexity, implementation cost, and total security. The key length of 64 bits, of which only 56 are actually used in the algorithm, achieves a high degree of security at modest link cost.

Critics of the DES, concerned with the choice of key length, have hypothesized a specialized decoding machine, consisting of 1 million parallel processors, which could break the code within 24 hours. It is estimated that such a machine could be designed and built for between $10 and 20 million, using presently available technology. In response, the supporters of DES argue that, if the user is concerned with this level of vulnerability, the data can be put through two sequential DES processes, which creates an effective key length of 112 bits, making brute force decryption impossible for several centuries.

The basic DES algorithm is shown in Figure 10-7. The encryption process operates on 64 bits of plain text at a time, combining the 64-bit key with 64 bits of the text. Each iteration, or computational cycle, 64 bits at a time, results in 64 bits of **ciphertext.** It is possible to simply perform a modulo 2 addition of the random key to the input text to create the ciphertext. However, because a single-stage

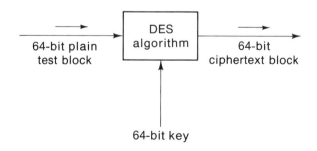

Figure 10-7.   Data encryption standard (DES) encoding.

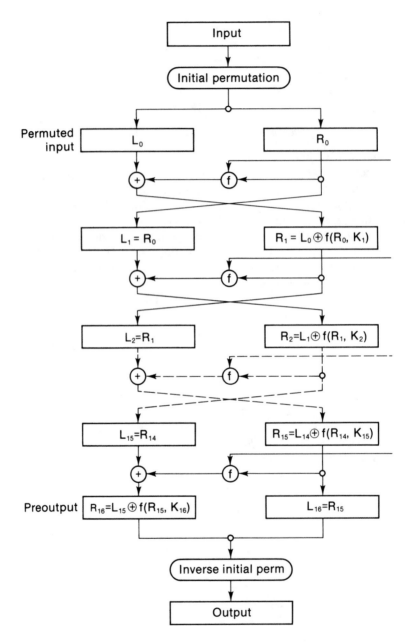

Figure 10-8.   DES internal algorithm.

encryption of two similar input blocks would produce similar output ciphertext, the result, in essence, would be a relatively simple substitution code that could be broken by intensive—but feasible—cryptoanalysis.

In order to insure that an exhaustive search is the only possible attack on the DES, the actual algorithm for encoding and decoding using the secret key is fairly complex, as shown in Figure 10-8. The 64-bit input block is initially permuted (scrambled) and then divided into two halves. The process then proceeds through 16 iterations, where each iteration combines the results of the previous iteration with a different subset of 48 of the 64 bits of the key. The standard shows the details of each of the operations, which consist of binary additions and bit shifts, as well as bit scrambling. The most significant points, for our purposes, are that the process results in ciphertext that is highly random and cannot be decoded except with the 64-bit key. In addition, the encoding/decoding process, being highly bit-oriented, can be readily reduced to implementation using large-scale integrated circuits. As a result, a number of vendors produce DES products in the form of complete encryption network elements or as device chips that can be incorporated into other network devices, modems, switches, and end terminals.

PRACTICAL APPROACHES TO SECURITY IN DISTRIBUTED NETWORKS

Achieving a practical level of security within distributed communications networks requires a combination of both technical and administrative control. Let us examine this combination of controls with reference to the seven-level ISO protocol hierarchy illustrated in Figure 9-13 of Chapter 9.

Protection from External Threats

In a public switched network, the external threats to the user information are primarily at the lowest levels of the interface hierarchy—at the physical, link, or network layer, where active or passive wiretapping or interception of the transmitted data is of greatest concern.

Figure 10-9 shows the potential application of DES to such a network, which provides end-to-end encryption for pairs of subscribers to the public network. The X.25 interface device at the user host's controls the operation of the external DES device. The DES process must pass without encoding the header information associated with each packet. Following the header, the first element of the level 4 protocol would be the encoding command that would initiate the encoding or decoding process on the remainder of the user information fields. The encoding process would then terminate at the end of the user data field with a similar command, followed by the link-applied error-check field, which would operate on the encrypted bit stream. In this way, the level 1, 2, and 3 protocol information would be unencrypted, and thus usable by the network for the control and routing

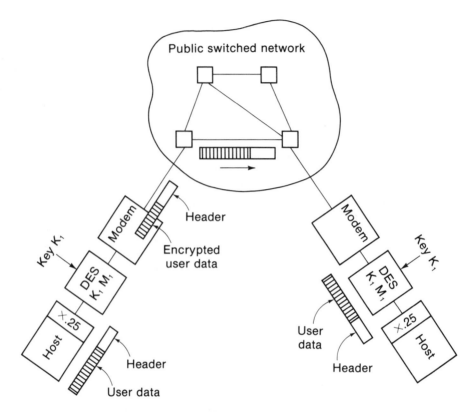

Figure 10-9.   Data field encoding using DES in a public switched network.

of the data, yet the user data itself would be protected from intelligible reception by an unauthorized device.

In a network of full-period connections, or a more centrally controlled network such as those typical of many local implementations, a more link-oriented approach is useful, as shown in Figure 10-10. Here, each link between the end users is protected individually by DES encryption devices, on either a dedicated or a shared basis. Since the protocol information is encrypted on the link along with the user data, the encryption process can be readily integrated into the modem or interface functions at levels 1 and 2 of the protocol hierarchy. Since all information is decoded at the central node, the node itself has to be well protected from intrusion. Aside from that vulnerability, the link-by-link approach is relatively simple to implement, provides each user with just a single key to deal with, and, as shown in Figure 10-10, permits very cost-efficient implementations because the DES encryption functions can be combined and integrated into the multiplexing function as long as users within the same community can share a common key.

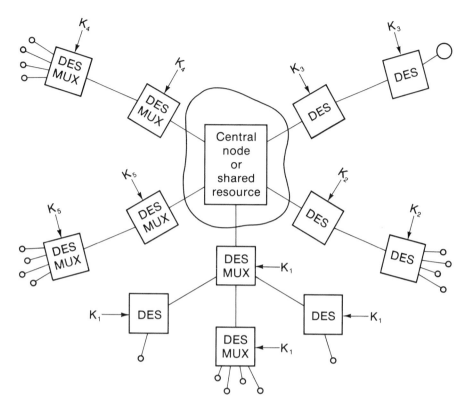

**Figure 10-10.   DES applied to a centralized or shared
resource network.**

Protection from Internal Threats

The most serious threats to many distributed networks and their host ADP
systems are from internal sources. Most known abuses or frauds perpetrated on
such systems have been carried out by either system insiders or persons with
intimate knowledge of the operational details of the system. In such systems,
administrative controls are essential. These kinds of controls, applied at the higher
levels of the protocol hierarchy, include such simple devices as user passwords,
file partitioning, and frequently run audit and check programs. Administrative
controls can be combined with technical facilities to even further enhance the
degree of protection.

For example, user access protection and file protection are often achieved
with passwords. However, it is not difficult for a system insider to access the
system, read the passwords for all users, and thus be able to masquerade as any
of the users. The DES encryption process can prevent this by simply encrypting
the password with a secret key and storing the encrypted version of the password

rather than the password itself. To access the system, an authorized user transmits his password. The system then encrypts the password with the secret key, using DES, and compares the result with the stored version of the encrypted password. If they match, access to the system is authorized. If an intruder "steals" the encoded password stored in the system and attempts to use it without knowledge of the secret key, access will be denied.

Similarly, DES coding can protect the files and information stored in the system by storing only protected data that has been processed by the DES algorithm before transmission. DES encoding of stored information can be applied to the processes and procedures needed to operate at the presentation and application layers of the protocol, so that the format as well as the content of the user information is transparent to the network operation.

Obviously, any of these protections can be defeated if the secret key or keys to the DES algorithm are compromised or "stolen." Here, administrative controls can periodically change the keys, for example, as it is far easier to reencrypt the password list than to make every user select a new password regularly.

SUMMARY

1. The security of distributed networks can be vulnerable to both external and internal threats because of the very nature of the technology.

2. The three main threats to networks are interception, insertion, and disruption of information.

3. Favorable trends in data processing costs favor protection based on the "combination lock" principle since the exponential relationship of decoding costs to encoding costs makes "brute force" attacks on a system unfeasible.

4. A number of different technical approaches can be used for network information protection, including link encryption, end-to-end encryption, and public key encryption. The widespread data encryption standard (DES), together with fundamental administrative protections, provides a high degree of economical protection for distributed communications networks.

SUGGESTED READING

BRYCE, HEATHER. "The NBS Data Encryption Standard: Products and Principles." *Mini-Micro Systems*, vol. 14, no. 3 (March 1981), pp. 111–116.

Bryce introduces the basic concepts of the NBS DES algorithm and the specific implementation of the algorithm in the Motorola product line. The article shows the application of these devices in a link-encrypted situation and indicates the flexibility of the product line by making the key components available for integration in other terminal or modem products.

CAMPBELL, CARL M. "Design and Specification of Cryptographic Capabilities." *IEEE Communications Society Magazine*, vol. 16, no. 6 (November 1978), pp. 15–19.

This paper provides a concise introduction to the use of cryptography for data secrecy, data authentication, and originator authentication in a communications environment. It distinguishes among link-by-link, node-by-node, and end-to-end protection, and shows the application of DES to each approach.

COHEN, KEN, and GARVEY, MICHAEL D. "Data-encryption Box Secures Comm Systems Easily." *Electronic Design*, vol. 29, no. 8 (April 16, 1981), pp. 159–163.

This article describes in detail the implementation of the DES algorithm provided by Western Digital Corporation with a single LSI circuit in combination with a few MSI interface circuits. The article shows the application of these devices to a DES encryption machine capable of operation at as high a link bit rate as 163,000 bits per second.

HELLMAN, MARTIN E. "An Overview of Public Key Cryptography." *IEEE Communications Society Magazine*, vol. 16, no. 6 (November 1978), pp. 24–32.

This excellent introduction to the concepts of public key cryptography provides several examples of the public key algorithms applied to small numbers to illustrate the principles. The article also includes a number of references to much more rigorous papers on public key cryptography.

U.S. Department of Commerce. "Data Encryption Standard." *Federal Information Processing Standards Publication 46*, January 15, 1977. Springfield, Va.: National Technical Information Service.

This publication defines the implementation, operation, and bit processing needed to protect data using the federally supported Data Encryption Standard. It shows the data processing that must take place in the encoding and decoding process, as well as giving an overall description of the background and applications of the standard.

11

Integrated Services Digital Networks

THIS CHAPTER:

will introduce the general principles of voice digitization and the handling of voice communications in a digital network.

will look at the techniques and advantages of voice compression and bandwidth reduction.

will show alternative ways that voice and data communications can be combined in distributed networks.

will present approaches to fully integrated communications services networks.

Most of our discussion so far has centered on the implementation of distributed data communications networks. Though voice communications were not specifically excluded from the discussions, the data rates and types of applications used in the discussions were clearly directed to the distributed data communications user.

This chapter will remedy that imbalance by introducing the various digitization approaches that allow voice signals to be compatibly placed on common distributed networks with data communications services. The combination of voice and data users into common networks is a rapidly evolving capability around the world. Though there are many different technical implementations, the general class of such integrated networks has come to be known as integrated services digital network, or **ISDN.**

PRINCIPLES OF VOICE DIGITIZATION

Because of economies in certain types of transmission media, voice signals have been digitized for transmission purposes for many years. These economies derived mainly from the fact that, if the voice signals were converted to a digital format, 24 voice channels could be simultaneously transmitted over a single pair of wires. For short-distance transmission (up to 50 miles or so), time-division multiplexing of digital voice signals was the cheapest way to utilize much of the wire-based transmission plant of the telephone networks of the late 1950s and early 1960s. In addition, the technology of voice digitization was pursued for security purposes since, as we saw in Chapter 10, the principles of encryption are applied to digital signals. It was much easier to protect voice communications converted to digital format than voice signals in their native analog format.

Pulse Code Modulation (PCM)

The most commonly used technique to convert analog voice signals into digital format is pulse code modulation (**PCM**). The basic PCM data rate of 64,000 bits per second per voice channel is based on the analog voice channel frequency bandwidth of 4000 Hertz, which requires a sampling rate of two times the basic bandwidth, or 8000 samples per second. Each sample is assigned one of 256 possible values, represented by an 8-bit binary code. The resultant data stream is thus 8000 samples per second, times 8 bits per sample, resulting in 64,000 bits per second. Figure 11-1 shows the process used in PCM transmission. For simplicity, this figure uses only eight levels, which could be encoded using 3 bits per sample. The analog waveform is sampled at discrete time intervals, with each sample being assigned a 3-bit codeword. The bits of the codewords are then transmitted as the sequence of ones and zeros over the binary, digital channel.

Digital PCM was originally applied to voice communications in order to multiplex 24 individual voice conversations onto a single pair of wires. The 8-bit-long codewords from 24 PCM channels are transmitted sequentially, resulting in a frame of 192 bits (8 bits times 24 channels). To each frame is added one additional bit to help keep the channel synchronized at each end, resulting in a transmitted frame length of 193 total bits. Since 8000 such frames are transmitted each second, the total channel rate of the familiar 24-channel PCM multiplex system is 1,544,000 bits per second (8000 × 193 = 1,544,000). This commonly quoted channel rate is the North American standard T1 channel rate. A European standard has also evolved over the years, based on a total of 32 channels, with an overall channel rate of 2,048,000 bits per second.

The early implementations of PCM achieved economy of scale by combining the analog channels and sharing the voice digitizer, in turn, among the 24 channels. In this case, the output digital stream is transmitted in binary format and processed through an inverse process at the receiving end of the circuit. The received data stream is converted back to sample values, filtered, and delivered to the output line. The multiplexing concept is illustrated schematically in Figure 11-2.

The digitization process is totally transparent and undetectable to the users since all the information in the originally transmitted voice signal is captured by the digital conversion process. Moreover, the digital signals suffer less degradation during the transmission process since noise cannot accumulate as the signal is repeated over numerous links.

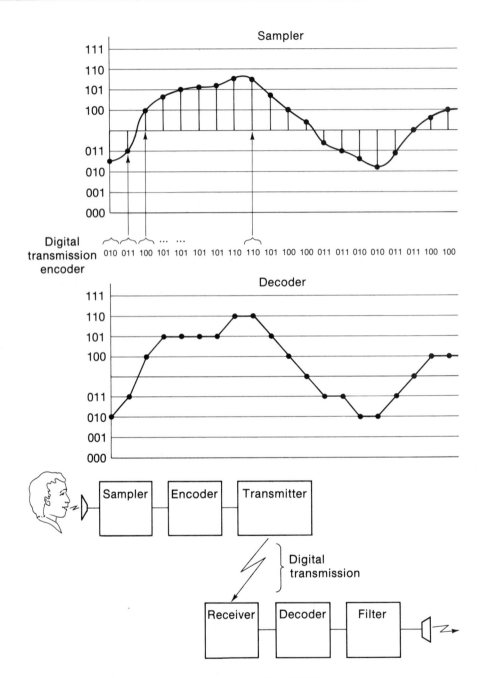

Figure 11-1.   The process used in PCM transmission.

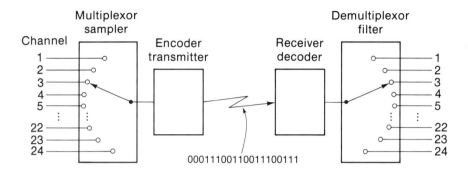

Figure 11-2. The multiplexing concept applied to PCM.

As the cost of digital processing hardware has been reduced, digitization of voice signals has become cost-effective for a broad range of transmission media and distances. More important, a number of new techniques have resulted in digital data stream rates considerably less than the 64,000 bits per second of PCM. However, because of the pervasive and widespread application of PCM in most common carrier facilities of the world networks, PCM is by far the most common form of voice digitization.

Differential PCM and Delta Modulation

By transmitting only the difference between the current sample value of the voice signal and the previous sample, a digitization technique called differential pulse code modulation (**DPCM**) can not only reduce the required bit rate but also improve the overall channel fidelity and quality. The principle of DPCM is illustrated in Figure 11-3. The improvement in the bit rate/quality tradeoff stems from the fact that the amplitude of the speech waveform changes at a relatively slow rate, and thus two successive samples only 125 microseconds (1/8000 of a second) apart will differ from each other very little. Sending only the amplitude differences each sample period requires fewer bits per sample for comparable accuracy of the reproduced signal at the receiving end, leading to channel bit rates of 32,000 to 48,000 bits per second, with quality as good as standard PCM at a rate of 64,000 bits per second.

Extending the concept of differential modulation results in a technique known as **delta modulation (DM)**. As shown in Figure 11-4, delta modulation again uses the fact that the speech waveform changes relatively slowly, and transmits only an indication that the current sample is larger or smaller than the previous sample. In order to faithfully reproduce the input waveform at the receiver, samples have to be taken more frequently than with PCM, but each sample is encoded into a single bit—for example, a binary one to indicate a larger value than the previous sample and a binary zero to indicate a smaller value.

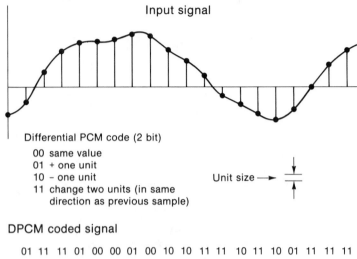

Differential PCM code (2 bit)

00 same value
01 + one unit
10 – one unit
11 change two units (in same
direction as previous sample)

Unit size ⟶

DPCM coded signal

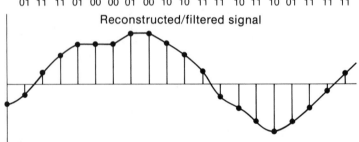

01 11 11 01 00 00 01 00 10 10 11 11 10 11 10 01 11 11 11

Figure 11-3.   Voice digitization with DPCM.

Taking the samples at a higher rate than 8000 times per second and encoding only the relative value in this way requires channel rates comparable to PCM for similar quality. However, adding a small amount of complexity, which increases the increment represented by each bit in proportion to the current rate of change of the speech signal, results in a technique known as continuously variable slope delta modulation (CVSDM, more typically called **CVSD**). CVSD encoders, operating at bit rates of 32,000 or 16,000 bits per second, achieve quality or performance comparable to PCM at 64,000 bits per second. Even more important, implementations of CVSD coders within a single large-scale integrated microcircuit permits economical encoding right at the user end instrument.

Analysis–Synthesis Techniques

PCM, DPCM, and CVSD all attempt to encode the input signal waveform in such a way that the waveform can be directly reconstructed from the encoded samples received at the destination. Another entire class of speech digitization

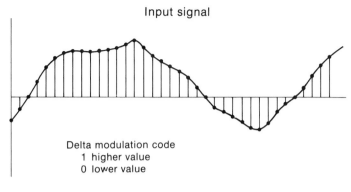

Input signal

Delta modulation code
1 higher value
0 lower value

Deltamod coded signal

1 1 1 1 1 1 1 0 1 0 1 1 1 1 0 0 0 0 0 0 0 0 0 0 0 0 0 0 1 1 1 1 1 1 1 1 1

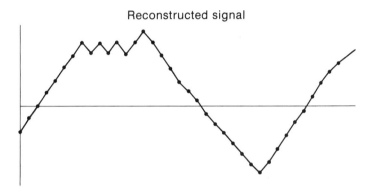

Reconstructed signal

Figure 11-4.   Voice digitization with delta modulation (DM).

techniques uses the principle of speech analysis, decomposition, and synthesis of the input signal. These analysis–synthesis techniques are based on the analytical representation of the human speech process, where each utterance is analyzed to determine the combination of parameters of the larynx, vocal cords, and nasal passage resonances that compose the current samples of the speech. Rather than transmitting the samples of the resulting spoken waveform, the encoder sends the parameters of the speech production process; a device at the receiver end of the link then synthesizes a replication of the input speech. Many fewer bits are required to accurately represent the speech parameters than to reproduce the original waveform since, because the channel input is human speech, a great deal of information is known in advance. As a result, excellent speech quality is achieved at data rates as low as 2400 bits per second. The processing devices at each end of the channel are much more expensive than PCM or CVSD devices, however— $1000 or more per terminal compared to about $10 per terminal for PCM or

CVSD—though recent innovations have reduced their implementation to three to six microprocessor-based chips. Nevertheless, the tremendous bit rate reduction makes these devices very desirable for many applications in spite of their cost. The most commonly available type of analysis–synthesis technique is known as linear predictive coding (**LPC**), although other techniques known as adaptive predictive coding (**APC**) and channel vocoders are also found in various applications.

The major disadvantage of analysis–synthesis techniques, aside from their cost, is the fact that they are based on the structure of the speech mechanism. Since the digitizers do not reconstruct the waveform from the transmitted samples directly, input signals to the encoder other than human speech cannot be transmitted accurately. Thus, modem signals or signals from, for example, a low-speed facsimile device, which could be readily sent over a PCM or CVSD digitized channel, cannot be used on an LPC channel. In practice, if the encoding process is done at the user-end terminal, this is not a serious problem. The encoding mechanism can be readily bypassed, and a baseband digital signal from a nonspeech device can be placed directly on the digital output line that would carry the encoded speech output. The difficulties are more significant in transitional stages of moving toward integrated services networks or when the speech encoder is shared on a centralized, switched, or trunk interface and tandem arrangement.

VOICE COMPRESSION AND BANDWIDTH REDUCTION

Applying digital processing to the conversion of speech into digital format offers the opportunity to reduce the total capacity required for transmission. The most direct approach to bandwidth reduction is using the lowest possible bit rate per voice channel carried. Thus, techniques like linear predictive coding, operating at 4800 bits per second, can carry 12 conversations in the bandwidth required by a single conversation using PCM at 64,000 bits per second. Since annual costs per voice channel are frequently many thousands of dollars, the ability to multiply capacity, even by using complex and costly processing devices at the terminals, is cost-effective in many cases.

Aside from the digitization process, the most direct approach to bandwidth compression results from the bidirectional characteristic of voice conversations. Since normal conversation consists of one party listening while the other is talking, half of the bidirectional circuit between two users is always idle. In addition, measurable gaps in speech—intersentence, interword, and intersyllable pauses—further reduce the average occupancy of each direction of the voice channel. It is possible to double the capacity of the transmission plant used in a network by sensing the activity on each half of the full-duplex circuits between network switches and using those circuits to carry only active transmission of information.

With a technique known as time assignment speech interpolation (**TASI**), the processing is carried out in special processors external to the network switching

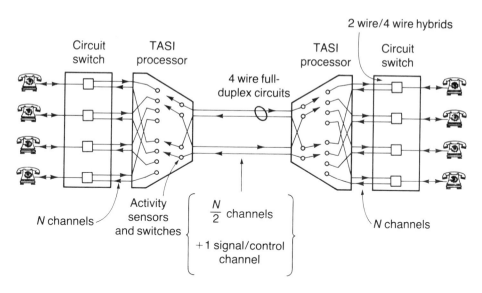

**Figure 11-5.   Implementation of time assignment speech
interpolation (TASI).**

devices, as shown in Figure 11-5. Typical applications derive two apparent
channels for each physical channel, most commonly 48 channels on a 24-channel
system; however, these same principles can be applied to smaller groups, such as
deriving 8 effective channels from 5 physical channels. As the number of channels
handled in a single system is increased, the probability of user lockout is decreased.

User lockout is a probabilistic situation that occurs if the momentary demand
of the current users exceeds the actual number of physical channels available. For
example, in a 48-channel system using 24 physical channels, lockout will occur
when 25 users are all trying to transmit in the same direction. This causes a
momentary delay for the 25th user until one of the other users pauses in his use of
the channel. As more and more channels are combined in such a system, lockout
is less likely.

Voice compression techniques can increase not only voice user capacity but
also nonvoice services, such as transmission of data packets when speech signals
are idle. Thus, various approaches to voice compression and bandwidth compression are widely applicable to voice and data integration.

VOICE AND DATA SERVICE INTEGRATION

Rapid advancement toward integrated services digital networks is motivated
by a combination of user service demand diversity and rapidly evolving technology.
On the network side of the user-network interface, service providers can dramatically increase total network traffic carrying capacity with little increase in facilities
by voice and data integration combined with digitized voice signals.

The TADI Approach

In time assignment data interpolation (**TADI**), an extension of the TASI concept, voice/data integration takes place on the interswitch trunk circuits. TADI uses the idle time periods in normal telephone speech patterns to insert bursts of data communications–based information.

Figure 11-6 shows a flexible implementation of the TADI technique using a common switch for voice and data signals. When the switch detects a pause in the activity on one of the circuit switched voice channels, it inserts a packet of data destined to the next switch. If the interswitch channel rate is based on a 64,000-bits per second digital voice channel, a 1000-bit packet will fit into a time period of about 15 milliseconds, which is less than the intersyllable gaps in normal speech. Clearly then, many packets can be interspersed into the gaps between words and sentences, and during listening periods, where the pauses range from one-half to many seconds. TADI thus provides the ability to carry very large amounts of data traffic with little or no increase in the transmission capacity of the basic circuit switched network.

TADI operates using packets or bursts of data buffered within the circuit switches. When the activity detectors on the interswitch trunks sense an absence

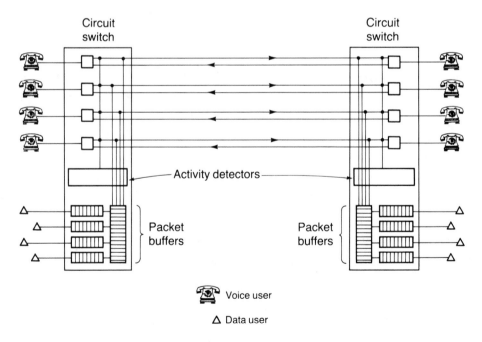

Figure 11-6.   Implementation of time assignment data interpolation (TADI).

of transmission, they signal the switch at the end of the circuit to quiet the circuit to the end user, and meanwhile insert data packets onto the circuit. When normal voice activity resumes on the circuit, the connection is returned to the circuit switched user within, at most, the length of a single packet, typically no longer than 15 milliseconds.

Voice and Data Integration via Packet Switching

The naturally bursty nature of processed, digitized speech can be used to integrate voice and data into a common packet switched network. In this case, the packet switched "connections" are arranged to emulate circuit switched connections through the network. Among the key characteristics of a circuit switched connection through a network is that it follows a fixed path, with constant delay through the network. It has a fixed overhead associated with the initial set-up of the call, but the overhead is independent of the length of the message or transaction. Thus, circuit switching is a more efficient mechanism than packet switching for very long data transmissions or for voice calls.

Figure 11-7 illustrates a typical configuration of a packet switched network, except that each switch contains a circuit reference table. Short data transactions proceed through the network using dynamic routing, flow control, and all the comprehensive processing features of packet switching. However, when the properties of a circuit switched connection are required, a call set-up protocol creates a direct route through the network via a set of table entries in each switch along the path. The logical connection through the network is referenced to the tables with a single logical connection number. Transmission proceeds using packet switching principles, except that each packet does not have to carry the full overhead normally associated with packet switching. After the initial call set-up, each packet needs to carry only the reference number to the table entry in each switch. When a packet arrives at a switch, the reference number is checked against the table, which indicates the proper routing and handling of the packet. Depending on the class of service provided, the packet would normally be put at the head of the queue on the proper outgoing line, thus keeping an essentially constant delay through the connection.

In addition, to fully simulate circuit switching, error checks would not normally be made on the packets designated as circuit switched, so no retransmission of packets would ever be necessary. Thus, a continuous stream of packets, with only a few bits of overhead in each packet, will flow through the network, over a path determined at call set-up time, and be held constant throughout the connection. For this approach to operate effectively, however, the emulated circuit switched connections would require continuous data rates that are only a small fraction of the overall capacity of the interswitch trunks. This approach provides the advantages of circuit and packet switching in a common network, implemented through protocol and software within a basically packet switched configuration.

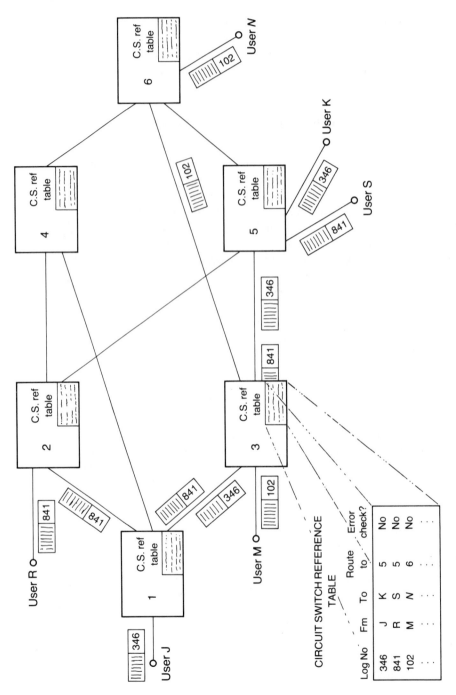

Figure 11-7. A packet switched network using circuit reference tables.

THE FULLY INTEGRATED COMMUNICATIONS
SERVICES NETWORK

Adding Image and Video Communications

In order to complete the integration of fundamental communications services into a single network, the ability to carry image and video communications must be added. By **image communications,** we mean the transmission of a still picture or motionless object. **Facsimile transmission** is presently the most common form of image transmission; an entire page of information is transmitted as is, rather than the digital representation of the letters and characters that were on the page. **Video transmission** adds motion to the image transmission; it can range from full-motion, full-color television signals requiring many megabits of communications capacity to freeze-frame video, a series of sequential still images transmitted over relatively low bit-rate facilities.

Basic Operation of an Integrated Network

The full complement of communications facilities would thus comprise voice, data, image, and video communications; these functions would be integrated into a network with an overall capability as depicted in Figure 11-8. Such an integrated services digital network has two essential features. The first is the fundamental conversion of all services into digital format, with the digital bit streams providing the "least common denominator" needed to achieve transmission plant integration. Second is the ability to employ intelligent processors, both at user concentrations and at network-user interface points to assure appropriate facility selection and bandwidth allocation.

Figure 11-8 shows separated circuit networks, packet networks, and other special-purpose networks. These are functional separations only; they can logically be drawn from a common pool of transmission media and facilities. As the common carrier networks evolve to a predominantly digital backbone plant, the network processing functions will be able to dynamically allocate portions of the total capacity to support an array of functionally independent networks and facilities, closely tailored to individual user characteristics.

Service Integration through User End Processing

The rapid proliferation of competing carrier services, combined with the relatively low cost of sophisticated computer-based processing functions, allows the network user a high degree of service integration. Service vendors will probably be able to fully integrate within five to ten years. Full integration of carrier-provided services, at least among the major long-haul carriers, is not likely until the late 1980s or 1990s, mainly because of the large sunk investment in older technology, as well as the magnitude of facilities replacement. Individual network users or user groups, however, can already achieve full service integration in a number of different ways.

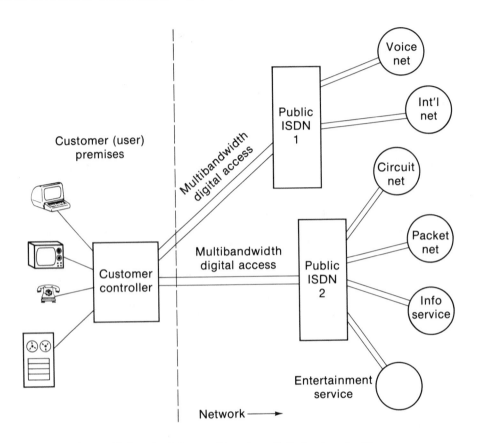

Figure 11-8.   An integrated services digital network. (Source: Dorros, 1981.)

The logical user device to achieve integrated services is a functional expansion of the present-day PBX, indicated in Figure 11-8 as the "customer controller." Current PBXs provide the customer control and interface to a wide variety of carrier-based services. These devices provide the intelligence to process, manage, and control user-premises devices; to convert signaling and format information among various protocols and services; and to select functions through routing and dynamic capacity allocation. Such devices enable users to evolve to an integrated, distributed communications architecture, as shown in Figure 11-9.

Architectural Features of the Integrated Network

The generic architecture shown in Figure 11-9 is essentially the summarization of the features and principles developed throughout the first ten chapters of this book. High-capacity, high-quality communications channels, largely over satellite facilities, provide the major long-haul component of the distributed

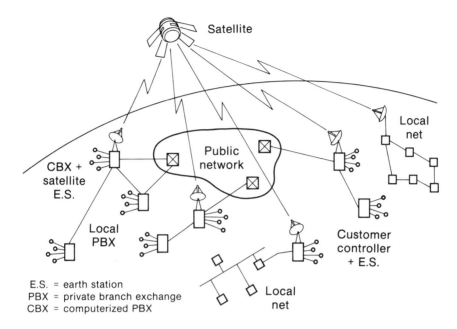

Figure 11-9.   An integrated distributed communications architecture.

architecture. Since the high-capacity, long-haul facilities are the most costly element of the communications network, they must be used on an efficient, demand-assigned basis. The burstiness of many of the user data sources entering such a network, even voice information, allows wide use of packetized, or packet-like, communications in the capacity allocation mechanisms.

Intelligent user end devices perform the local distribution of the connectivity and capacity with a variety of techniques. In general, baseband techniques are best suited to low-cost data communications users; more expensive, but more capable, broadband techniques can be employed for the full range of integrated services, including video and voice.

The user-end switching device, or customer controller, can vary greatly in complexity and functions. It can serve as the integrating vehicle, and at the same time provide the optimum processing necessary to effectively manage the many different user services accessible to the network. As a minimum, the customer controller embodied in a basic PBX must select among a number of different transmission media and routes, usually from a variety of suppliers, in order to permit the interconnection of two customers within the network. It also will provide the integration mechanism for the different user end services.

The Customer Controller—A Range of Possibilities

There are two considerably different approaches to employing the customer controller or enhanced PBX as the integrating vehicle for a distributed network.

Integration Based on Voice Channels. In the most widely used approach, the voice users and voice channels provide the fundamental building block of the network, and nonvoice services are transformed into voicelike or voice-equivalent channels. The basic structure is illustrated in Figure 11-10.

This approach permits the user end instruments to remain analog for voice service, yet local distribution of connectivity for data users can be on a direct-wire, digital baseband facility. Data users within the local area are connected through the customer controller without the need to convert the baseband digital signals to voiceband equivalent parameters. When a data user needs connectivity outside the local area, the connection is completed through a modem pool, which allows one of a number of possible modems to be connected to that user. This approach leads to significant savings in total equipment since, unlike the permanent assignment of a modem or data set to every user in the local area, many fewer modems are generally needed in the modem pool than the total possible digital end users. For many users, integration lies in the ability to use the same connection to the PBX alternately for either voice or data communication functions, although both generally cannot be employed at the same time. Connection to the long-haul portion of the network is based on voiceband analog channels, or possibly on a wideband digital format.

More capable versions of this approach to integration will add the ability to multiplex low-speed data devices onto common wider bandwidth channels. In

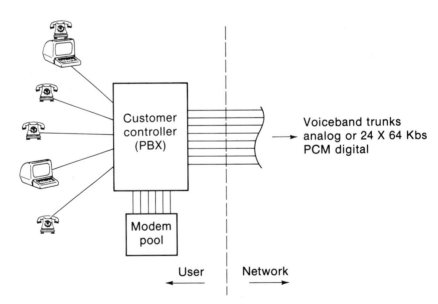

Figure 11-10.   Integration through a customer controller based on voice channels.

addition to saving modem devices, this reduces the total number of switch connections needed to service the total user community and permits more efficient use of the long-haul lines in the system. In some systems, the PBX has the further ability to treat the digitally derived interswitch facilities differently from the basic voiceband analog derived channels. They are thus able, for example, to handle many more low-speed users in a single 64,000-bits per second voiceband-equivalent channel than in the 4 KHz analog equivalent.

Integration Using Digital Circuit Switching.   Another approach to integration is digital time-division circuit switching, as illustrated in Figure 11-11. All inputs and outputs to the customer controller or PBX are digital, thus requiring that voice user devices be digitized at the instrument itself. Input lines to the switch from the digitized voice instruments can be used alternately for voice services or data services; some implementations can combine voice and moderate-speed data services simultaneously. For example, a single 64,000-bits per second digital port on the PBX could handle a 56,000-bits per second voice user at the same time as a data port with an average input rate of 8000 bits per second or less. Nonvoice ports on the switch are served through a multiplexor, improving the switching efficiency of the potentially large number of low-speed data devices in such an installation. Trunk side interconnections are fundamentally digital, either to channel interfaces operating at multiples of the basic T1 channel rate (e.g. 6.3 Mb/s) or into the digital baseband of the satellite earth stations or long-haul transmission plant.

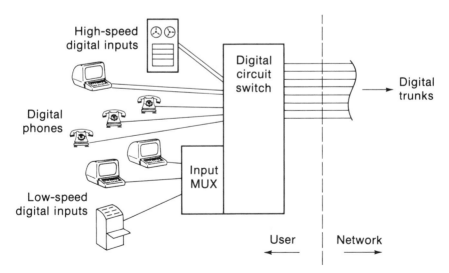

Figure 11-11.   Integration using digital circuit switching.

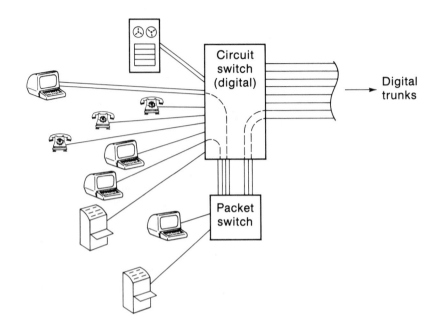

Figure 11-12. Integration by combining circuit and packet switching.

Integration by Combining Circuit and Packet Switching. Another approach to integration combines voice-based circuit switching with packet switching (Figure 11-12). The circuit switch portion may be analog voiceband based or fundamentally digital, depending on the economics of the specific hardware design. The packet switching function could be software derived within the basic switch, or it could be entirely self-contained as an external device slaved to the customer controller. Voice users would be handled directly by the circuit switching portion. Data users could be interfaced directly to the packet switching portion or could access it indirectly through the circuit switching matrix.

This approach offers the advantage of high capacity and low overhead for voice and high-volume data users, while providing local access to the features and processing of the packet switch. Since the packet switches share their trunking with the rest of the integrated network through capacity assignment using the circuit switch, the overall capacity of the network can be managed to insure low delays for the real-time data that is generally predominant in a packet switching environment.

Integration Using Packetized Voice and Data. Yet another approach to integration uses packetized voice together with the other communications modes in a common packet network. Figure 11-13 shows that the voice digitization process takes place

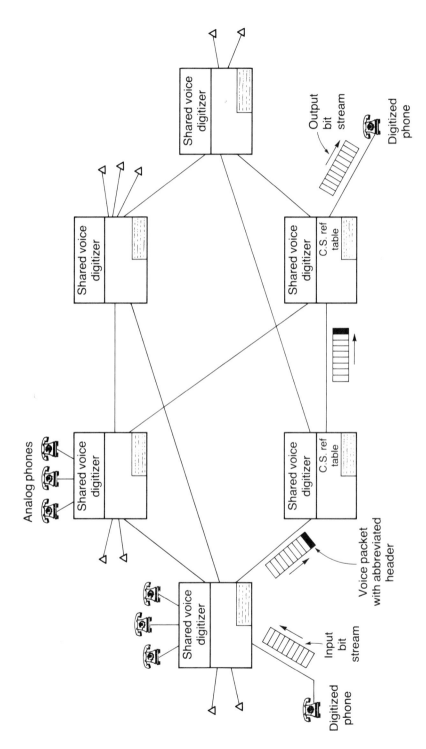

Figure 11-13.   Integration using packetized voice and data.

at either the user end instrument or the switch input. If the voice processors are placed at a concentration point, such as the switch input, their cost can be shared among many more users since the processors can be pooled to handle the number of active lines at the switch and not the total number of end users. Packetizing the voice at the end instrument makes better use of the input facilities and allows any given user to handle voice and data communications simultaneously over a single connection into the switch.

In a purely packet switched approach to integration, the voice packets have to be class marked so they can be handled as expeditiously as possible. Voice packets need not be error checked since there is no time to retransmit errored packets. Packets received with errors will be processed at the output speech synthesizer; the worst result will be a short noise burst if the errors are severe. Since voice packets will contain a relatively small number of bits, it is important to minimize their overhead. This is done by establishing a fixed path through the network during the call set-up and transmitting all packets associated with that call over the same path. A table entry in the memory of each switch along the end-to-end path associates the packets with a logical circuit, so the only overhead required on each packet after the call set-up is the logical circuit reference number. Each switch can then reference this number to the table entry, which instructs the switch on the handling and routing of the packet.

NEAR-TERM TRANSITION TO INTEGRATED SERVICES DIGITAL NETWORKS

The implementation of integrated services digital networks is greatly facilitated by the advanced capabilities of many new local area networking devices, such as cable systems, PBXs, and radio-based distribution systems. However, a large percentage of the common carrier plant must be converted to digital service. Current plans indicate that more than 60% of the Bell System switching and transmission plant will become digital by 1985. It is thus significant that the carriers are planning to offer end-to-end digital services combining high bandwidth analog facilities and digital trunking facilities based on the 56,000/64,000-bits per second digital data rate.

Figure 11-14 shows the transitional approach likely to be used within the public telephone plant. At least over relatively short distances, the metallic local loops and the metallic crosspoints of the space division matrices of the No. 1A electronic switching system (1A ESS) provide the bandwidth necessary to access the digital toll backbone. The toll backbone operates within the 64,000-bits per second, digital PCM hierarchy via the Number 4 ESS and T-Carrier transmission facilities. Analog voice services are converted to digital at the toll trunk interfaces.

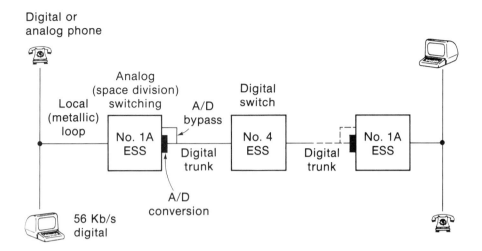

Figure 11-14.   Transition of the public telephone plant to digital operation. (Source: Johnston and Litofsky, 1982.)

Digital terminals will operate at rates of 56,000 bits per second, be carried directly over the metallic parts of the connection, and be synchronized into the digital toll backbone. As the local plant is upgraded to digital operation, the metallic portions of the connection will be upgraded with both digital transmission and switching facilities. This type of operation will facilitate direct interface with many of the local networking concepts; for example, locally connected packet switching processors will be able to readily derive direct processor-to-processor connections at 56,000 bits per second with economical switched connections on a demand/usage basis.

The technology and marketplace can very effectively combine bursty format (packetlike) data transmission with other services in digital format in common networks. Such data can be handled over a wide range of facilities, from local wirelines to long-haul satellites. Though absolute costs for such services may rise, the costs of communications relative to other services will continue to decline because of the technology as well as the economies of scale that result from the combination of many services into integrated networks. Consequently, the network user has to look beyond his immediate needs to be able to accommodate the many likely distributed communications services that will evolve.

SUMMARY

1. Distributed communications networks that meet the needs of geographically distributed data users can be expanded to include additional communications sources.

2. Voice, video, and graphical image sources converted to digital format are implemented in many different ways in integrated services digital networks (ISDNs).

3. Voice signals can be converted to digital format using PCM, CVSD, LPC, and other techniques, with bit rates between 64,000 BPS and 2400 BPS per voice channel. In general, higher complexity and greater cost are needed for lower bit rates.

4. Once digitized, voice signals can be integrated with data sources by many different switching techniques, with digital circuit switching and packet switching providing the most promising capabilities.

5. Most distributed networks will evolve with a combination of satellite long-haul and digital local facilities.

SUGGESTED READING

DORROS, IRWIN. "ISDN—A Challenge and Opportunity for the 80's." *IEEE Communications Magazine*, vol. 19, no. 2 (March 1981), pp. 16–19.

This article introduces the Bell System approach to the integrated services digital network. The author, Assistant Vice President for Network Planning for AT & T, shows the evolutionary approach being taken by AT & T as a result of the natural modernization and upgrading of the public network to digital facilities.

FRANK, HOWARD, and GITMAN, ISRAEL. "Economic Analysis of Integrated Voice and Data Networks: A Case Study." *Proceedings of the IEEE*, vol. 66, no. 11 (November 1978), pp. 1549–1570.

This paper reports on a major study performed for the U.S. Department of Defense comparing packet switching to other techniques for general applicability to mixed user networks. A full range of network switching technologies is studied. Summarizing both the technical and the economic parameters considered in the analysis, the article shows that packet switching provides the most cost-effective topology under a wide range of assumptions.

JOHNSTON, STANLEY W., and LITOFSKY, BARRY. "End-to-End 56 KBPS Switched Digital Connections in the Stored Program Controlled Network." *Business Communications Review*, vol. 12, no. 1 (January–February 1982), pp. 17–23.

Two engineers from the Bell Telephone Laboratories describe the near-term capabilities of the hybrid analog-digital public network, which permits the transmission of end-to-end digital services at 56,000 bits per second. In conjunction with intelligent switching and end-user devices, the capability provides the initial steps along the path to longer-range integrated services digital networks.

OCCHIOGROSSO, BENEDICT. "Digitized Voice Comes of Age," Parts 1 and 2. *Data Communications*. vol. 7, nos. 3 and 4 (March and April 1978).

This two-part article describes the currently available techniques for converting voice signals into digital format. The basic technologies are explained, showing the different bit rates, performance characteristics, costs, and tradeoffs associated with each. Sources of voice-processing hardware are described.

12

Design of a Nationwide
Distributed Network: A Case Study

THIS CHAPTER:

will illustrate the design of a hypothetical nationwide network
combining satellite and local network connectivity.

will show the relationship among the various elements of the
network that influence the optimization of the design.

will show how the costs of key components, such as earth
stations, greatly influence the resulting designs.

System concepts for large communications systems have traditionally posited a backbone network with a user-defined access line connectivity to the backbone. Satellites have been considered a high-quality, highly reliable source of very long-haul circuits, and viewed as "cables in the sky." We have examined numerous techniques that permit satellites to be used much more intelligently and flexibly than in a simple point-to-point manner. In handling users on a distributed network basis, the satellites take the principal role in all network connectivity, including most of the access network, in addition to the main backbone. If this concept is carried to its natural conclusion, the network backbone is eliminated entirely. In this chapter we will illustrate the design of a hypothetical communications network that combines satellite and local connectivity.

A LARGE, NATIONWIDE USER NETWORK

The user community for this conceptual design study will encompass a diverse set of about 2500 geographically dispersed user concentrations. In the absence of a satellite system the users would be connected by leased terrestrial access lines between themselves and large backbone switches, using leased terrestrial trunks between the switches. If a satellite system is used, the costs of the leased terrestrial facilities can be avoided. A cost optimization based on the number of earth stations results from the fact that, although a large number of earth stations increase the satellite system cost, they minimize the terrestrial component costs. Too few earth stations lead to inefficient use of the satellite resources and require excessive, costly terrestrial interconnects. Since earth stations tend to be costly, fixed capital investments, their number must be carefully chosen to balance the cost factors.

This design example is aimed at the situation where a large private network user could potentially fulfill most of his capacity requirements with satellite facilities. It is also a feasible model where a new domestic satellite carrier, with a charter to serve users anywhere in the country, balances the cost of the earth stations access and terrestrial connectivity.

Overall Network Concept

Our overall network concept is based on current and future classes of high-technology, high-capability satellites having sufficient bandwidth and transmitter power to support a large number of relatively small, simple, and inexpensive ground terminals. The satellites and earth stations would constitute the substance of the communications network rather than providing supplemental capacity to meet special or long-haul requirements.

The degree to which terrestrial facilities would be replaced would naturally be governed by the comparative technical capabilities and economics of the satellite network components. In an ideal case, if the number of satellite ground stations employed was the same as the number of backbone switches, then the entire terrestrial backbone would be eliminated. If, then, the number of earth stations could be economically increased by the number of access areas—that is, user concentrations—the entire terrestrial network, both backbone and access, could be eliminated. Individual satellite ground stations would perform the concentration, switching, and transmission functions, and the need for backbone switches would disappear. In practice, we would expect the optimum level to fall between these two extremes, with earth stations placed at many, but not all, user locations.

The User Model

The model begins with a set of known user locations, the concentration profile of those locations, and a relatively well-designed, conventional, nondistributed backbone network capable of meeting the user communications demands. The model combines both the voice and the data requirements of the user community. The overall design is simplified by grouping and averaging the demands into various categories, rather than treating each user location individually.

Although design concepts such as this use a great deal of data averaging and smoothing, they are sufficiently adequate for system optimization studies and conceptual development of the networks to be deployed. Detailed design and implementation of the selected configuration would of course require more detailed, site-by-site study. In addition, since the basic traffic parameters are, at best, gross estimates, the designs tend to be highly conservative. The actual design is refined over time as operational experience with the network is obtained.

Since an efficient network design was known, employing 70 terrestrial switching nodes, this study starts with 70 rectangular service areas, centered around each of the terrestrial switches. The overall continental United States is thus approximated by a rectangular service area, 1500 miles by 2700 miles, yielding each of the 70 service area's dimensions of 215 miles by 270 miles. User access areas or user clusters within each service cell are distributed approximately uniformly throughout the cells. Traffic flow is distributed uniformly among the 70 cells, and, within each cell, traffic to each user location is proportionate to the number of user personnel at each facility. The use of 70 service cells as the starting point in this model is simply a matter of convenience; any number could actually be used. One cell for each major city or metropolitan area to be served might be chosen, for instance. On the other hand, a network might be known using a specific number of initial satellite earth station locations, and the model extended from

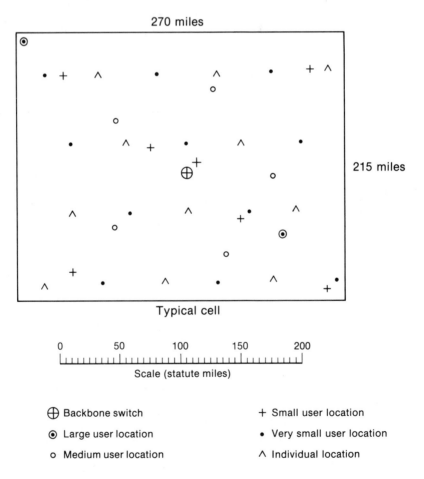

Figure 12-1.   User access area geographical distribution.

that point to determine the number of additional earth stations needed to optimize the design.

Figure 12-1 shows the uniform distribution of the user community within the rectangular service area. The salient features of the user requirement profile in this study are given in Table 12-1. Five different categories of user locations are shown: individual locations with a single user; very small locations, with an average of

Table 12-1.   Satellite Study Access Area Model

Access area size category	Large	Medium	Small	Very small	Individual	Total
Number of locations per cell	2	5	7	11	11	36
Total locations in U.S.	140	350	490	770	770	2,520
VOICE COMMUNICATIONS						
Voice terminals per location	1,835	346	34	6	1	
Access line traffic (erlangs)	74	13.84	1.36	0.24	0.20	
Access lines required per location; GOS = P.05	80	19	4	2	1	
Average distance to switch (miles)	135	70	90	90	88	105
Total lines in U.S.	11,200	6,650	1,960	1,540	770	22,120
Total channel-miles in U.S.	1,500,000	465,000	176,000	138,500	67,600	2,347,100
DATA COMMUNICATIONS						
Data traffic in busy hour (per location kilobits)	103,530	33,670	8,833	1,516	418	
Equivalent load in erlangs @ 16 Kb/s lines	3.6	1.17	0.31	0.05	0.01	
Additional lines per location; GOS = P.05	7	4	2	1	1	
Total additional lines in U.S.	980	1,400	980	770	770	4,900
Total additional channel-miles in U.S.	132,000	98,000	88,000	68,000	67,600	453,600
Total lines in U.S.	12,180	8,050	2,940	2,310	1,540	27,020
Total channel-miles in U.S.	1,632,000	563,000	264,000	206,500	135,200	2,800,700

6 users; small locations, with an average of 34 users; medium locations, with an average of 346 users; and large locations, such as corporate headquarters, industrial parks, regional headquarters, distribution centers, and the like, with an average of 1835 users. The data communications traffic is estimated separately from the voice traffic, with the capacity needed to handle the data traffic added directly to the capacity needed for the voice traffic. This provides an upper bound to the capacity requirement (and thus network cost) since we would expect our integrated network concepts to use capacity much more efficiently, resulting in lower demands for total capacity. A grade of service (GOS) of .05 (5%) is used for sizing the line requirements; in other words, one out of twenty call attempts would be blocked during the busy hour of the day, a highly acceptable network grade of service in commercial practice.

METHODOLOGY FOR NETWORK COST OPTIMIZATION

Basis of Estimation

A cost optimization analysis was performed on this model of the user environment to determine if there was a concave relationship for the total transmission system cost as a function of the number of satellite earth stations. Such a relationship would insure that system cost minimization could be achieved by selecting the proper number of satellite earth stations.

As the number of earth stations in the network increases, the dominant portion of the total system cost becomes that of the ground portion of the satellite system. Recognizing this, and in order to insure feasibility of the satellite-based operational concept, very conservative estimates of the cost, performance, and capacity of the space segment of the network were used. Costs were deliberately overestimated and capacities were underestimated, resulting in a satellite space segment that is more costly and has more capacity than is really needed. In addition, the space segment of the system was increased in size and complexity as the number of ground stations was increased, since the satellite downlink power would have to be increased to serve a larger number and lower performance set of earth stations. Table 12-2 indicates the nature of the space segment for different ranges of number of ground stations supported. As the number of ground stations increases, the satellites are presumed to require additional weight to permit more sophisticated beam steering and switching of narrow-beam antennas, and additional power to service the smaller ground stations. The estimated satellite cost, as well as the launch costs, are shown, with launches assumed to be in pairs of satellites.

Several quantities of ground stations are of particular significance here. Referring to the basic user model, the quantities of 70, 210, 560, and 1050 ground stations represent, respectively, ground stations at all backbone switches; all large locations and backbone switches; all medium, large, and backbone switch loca-

**Table 12-2. Satellite Space Segment for Various Ground
Station Configurations**

Ground Stations	No. of Satellites	Est. Sat. Weight (#)	Est. Cost ($ Mil) per Sat	No. of Launches	Cost for ($ Mil) Launch
Up to 50	2	1500	15	1	22
50 to 100	4	1500	15	2	22
101 to 350	4	2200	20	2	26
351 to 1050	6	2200	20	3	26

tions; and all small, medium, large, and backbone switch locations. Attempting to extend satellite earth stations to the very small and individual locations of the user model proved ineffective in reducing system costs. However, recent progress in reducing earth station cost, particularly in conjunction with the low-cost television receive-only (**TVRO**) satellite earth stations, may change this situation within the next few years.

Line Rates and Other Cost Parameters

A uniform channel rate of 16 kilobits per second (Kb/s) was used in order to simplify the computations. The 16 Kb/s rate provides acceptable voice quality for digitized voice, at very low cost, using a number of different techniques. For example, the Satellite Business Systems (SBS) satellite network uses 32 Kb/s voice channel rates, with 2:1 voice compression/activity detection to achieve an effective rate of 16 Kb/s per voice channel. In addition, 16 Kb/s provides sufficient data transmission bandwidths for most ADP applications. Though lower voice digitization rates would significantly reduce the cost of the leased terrestrial facilities, operating at data rates significantly less than 16 Kb/s would also reduce the power and data rate requirements (and hence costs) for the satellite system. However, the terrestrial costs most likely would reduce more rapidly than the satellite costs, thus narrowing the advantage of the satellite system employing many ground stations. This effect would decrease the optimum number of earth stations but would not change the fundamental conclusions of our design model.

All cost estimates were done using a ten-year system lifetime, a ten-year present-worth analysis, and a 10% capital cost recovery factor. Terrestrial trunk circuits (high density, high capacity) among switches were assumed to have a nominal monthly cost of $1.00 per mile, and access line circuits (low density, small cross-sections) were assumed to cost $2.00 per mile per month. All terrestrial circuits were assumed to have a fixed monthly distance-independent cost of $175 per channel end. The resulting ten-year present-worth costs were about $110 per mile per access circuit and $55 per mile per trunk circuit, plus about $13,750 per circuit end in fixed costs.

The Satellite Subsystem

The satellite system components were assumed to consist of a class of technologically advanced satellites and three different size categories of earth stations. The satellites would operate in a time-division, multiple access (TDMA) demand-assignment mode. Where a large number of earth stations (50 or more) are employed, the satellites would use a number of steerable, switchable, narrow-beam antennas. While sufficient capacity appears to be available in the 6/4 gigahertz (GHz) commercial or 8/7 GHz government microwave bands to handle the traffic for all cases, it would probably be desirable to utilize the 14/12 or 30/20 GHz region of the spectrum for much of the peak load, especially for some of the larger ground stations where antenna gain and available power could maintain

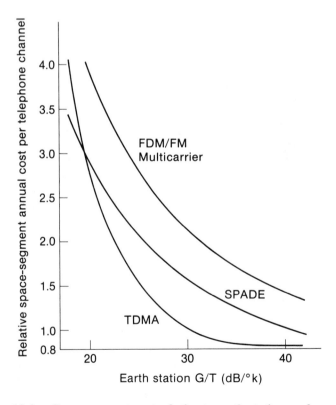

Figure 12-2.   Space segment cost relative to earth station performance.

communications in all but the worst of rainstorms. In addition, use of the high-frequency bands would permit the deployment of small earth stations, which would probably prove cost-effective at many of the very small and individual locations not included in the ultimate range of results in this analysis.

The ground stations are categorized as large, medium, and small, with performance ratings (gain to temperature, or G/T) of 40, 30, and 18 dB, respectively. This corresponds approximately to 60-, 18-, and 8-foot diameter antennas operating at the lower frequency bands (below 8 GHz). We can use this information, together with the information shown in Figures 12-2 and 12-3, to estimate the relationships between satellite system size and costs.

Figure 12-2 shows the approximate relationship between the performance of the earth stations and the cost of the space segment per voice channel. As the earth station gets larger, resulting in higher values of G/T, the amount of satellite power and antenna gain used per voice channel decreases, resulting in lower relative cost per channel. This effect really results from the fact that many more voice channels can operate simultaneously, over a single satellite transponder, when the access to the satellite is via a large, powerful earth station. Figure 12-2 also shows

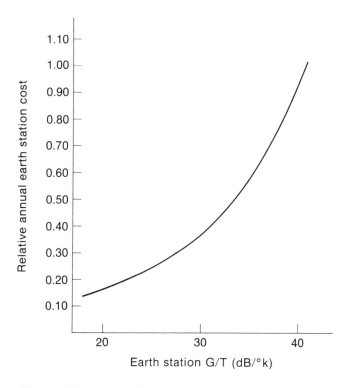

Figure 12-3.   Earth station cost relative to performance.

the effects of different forms of modulation and satellite capacity division, including traditional frequency division multiplexing (**FDM/FM**), single digital channel per carrier, demand assignment (**SPADE**), and time-division multiplexed, demand assignment (**TDMA/DA**). Because of its purely digital operation and high permissible peak powers, the TDMA/DA is by far the most cost-effective with respect to satellite utilization.

Figure 12-3 shows the approximate annualized costs of earth stations as a function of their performance factors. Note the reverse trend of this curve compared to the curves of Figure 12-2. Largely because of the rapid increase in size, larger antennas being required to achieve larger values of G/T performance, the installation, operation, and capital recovery costs of higher performance earth stations increase rapidly.

From these two figures we see, for example, that the relative space segment cost using a 30 dB G/T ground station is about 20% higher than the cost using a 40 dB G/T station using TDMA/DA. On the other hand, the cost of a 40 dB G/T earth station is about 200% higher than that of a 30 dB G/T station.

In satellite systems employing a relatively small number of earth stations, the overall costs are space segment dominated, so that saving 20% of the space segment costs far outweighs the effects of the more expensive earth terminals. However, in this analysis, as the number of earth stations increases, the costs become more dominated by the earth segment costs, thus making it desirable to use smaller earth stations at the expense of a more costly space segment.

As a result of these considerations, analysis of cases using 50 or fewer earth stations utilized only large earth stations, with performance figures of 40 dB G/T. For cases between 51 and 560 earth stations, only medium (30 dB G/T) earth stations were used. For cases with more than 560 earth stations, medium-size earth stations were used to serve the backbone switches and large and medium user access areas, while small (18 dB G/T) earth stations were used to serve the small access areas. Since the unit costs of the earth stations are subject to wide variations, owing to both improving technology and local variations in installation and engineering, the results of this analysis are derived parametrically, with the earth station unit costs being one of the key varying parameters.

CALCULATING OPTIMUM EARTH STATION POPULATION

By using the cost factors and methodology described above, we can derive cost curves that relate the transmission system cost associated with the user model to the number of satellite earth stations in the overall system design. The baseline costs for an all-terrestrial implementation were computed, using the relationship that:

$$C_0 = C_t + C_a$$

or

$$C_0 = m_t C_{tm} + 2n_t C_{tt} + m_a C_{am} + n_a C_{at}$$

where:

C_0 = system cost with 0 ground stations
(i.e., all-terrestrial system)

C_t = trunking costs

C_a = access costs

m_t = trunk channel-miles

m_a = access channel-miles

C_{tm} = cost per channel-mile for trunks

C_{am} = cost per channel-mile for access lines

n_t = number of trunk channels

n_a = number of access channels

C_{tt} = termination cost per trunk end

C_{at} = termination cost per access end

For the all-terrestrial system design, the relevant parameters, as previously indicated in Table 12-1 and the text, are:

m_t = 5,000,000 channel-miles

m_a = 2,800,700 channel-miles

C_{tm} = $55 per mile (10-year present worth)

C_{am} = $110 per mile (10-year present worth)

n_t = 7800 trunks

n_a = 27,020 access lines

$C_{tt} = C_{at}$ = $13,750 per termination

Substituting these values into the cost equation results in:

$$C_0 = 5 \times 10^6(55) + 2(7800)(13,750) + 2.8 \times 10^6(110)$$
$$+ \ 27,020(13,750)$$

C_0 = 1.169×10^9 for a ten-year life cycle, which is a total system
baseline cost of about $117 million per year

For the system configurations that employ satellites, the general expression for the system cost is:

$$C_N = C_0 + C_S - C_{AN}$$

where:

N = number of earth stations

C_S = satellite subsystem cost

C_{AN} = terrestrial system cost avoidance as a result of the satellite
subsystem

The cost of the satellite subsystem is derived from the sum of the launch costs, the space segment cost, and the aggregate cost of the ground stations, including both acquisition and operation and maintenance costs. The space segment and launch cost estimates can be estimated directly from Table 12-2 for various values of N, the number of earth stations. Defining C_{ss} as the space segment costs, we find:

$$C_{ss} = \$52 \times 10^6 \qquad 2 \leq N \leq 50$$
$$C_{ss} = \$104 \times 10^6 \qquad 50 \leq N \leq 100$$
$$C_{SS} = \$132 \times 10^6 \qquad 100 \leq N \leq 350$$
$$C_{SS} = \$198 \times 10^6 \qquad 350 \leq N \leq 1050$$

The ground station costs, based on the number, size, and unit cost of ground stations, are estimated as:

$$C_{GS} = 1.65NC_g$$

where:

$$C_{GS} = \text{ground segment ten-year cost}$$
$$N = \text{number of ground stations}$$
$$C_g = \text{ground station unit cost}$$

The factor 1.65 allows 10% of the original ground station cost for annual operation and maintenance expense, computed over a ten-year lifetime, present-worth analysis. Because different sized earth stations have to be used over different ranges of the number of earth stations, the ground station cost equation depends on the range of values of N. Recall that, for systems of fewer than 50 earth stations, only large earth stations are used. In the range between 50 and 560 earth stations, only medium-sized earth stations are used, and when more than 560 earth stations are used, all the earth stations greater than the first 560 are small. The first 560 earth stations used, which are located at the switches as well as at large and medium user concentrations, are all medium-sized earth stations. When earth stations are deployed to small user concentrations, small (18 dB G/T) earth stations are utilized.

This results in ground segment costs over the ranges of interest of:

$$C_{GS} = 1.65NC_1 \qquad\qquad\qquad \text{for} \quad 0 \leq N \leq 50$$
$$C_{GS} = 1.65NC_m \qquad\qquad\qquad \text{for} \quad 50 \leq N \leq 560$$
$$C_{GS} = 1.65(560)C_m + (N - 560)C_s \quad \text{for} \quad 560 \leq N \leq 1050$$

where:

$$C_1 = \text{purchase cost of a large (40 dB G/T) earth station}$$
$$C_m = \text{purchase cost of a medium (30 dB G/T) earth station}$$
$$C_s = \text{purchase cost of a small (18 dB G/T) earth station}$$

The total satellite subsystem cost, C_S, is thus the sum of the space segment, C_{SS}, and the ground segment, C_{GS}, costs, combined over the various ranges of interest.

Thus:

$$C_S = 52 \times 10^6 + 1.65NC_1 \qquad \text{for} \qquad 2 \leq N \leq 50$$

$$C_S = 104 \times 10^6 + 1.65NC_m \qquad \text{for} \qquad 50 \leq N \leq 100$$

$$C_S = 132 \times 10^6 + 1.65NC_m \qquad \text{for} \qquad 100 \leq N \leq 350$$

$$C_S = 198 \times 10^6 + 1.65NC_m \qquad \text{for} \qquad 360 \leq N \leq 560$$

$$C_S = 198 \times 10^6 + 1.65(560)C_m \qquad \text{for} \qquad 560 \leq N \leq 1050$$
$$+ \, 1.65(N - 560)C_s$$

Terrestrial Network Cost Avoidance

Having computed the baseline terrestrial network costs (C_0) and the satellite subsystem costs (C_S), it is necessary to estimate the terrestrial network cost avoidance, based on the satellite earth station deployment, so that the expression for total network cost, $C_N = C_0 + C_S - C_{AN}$, can be evaluated.

The terrestrial system cost avoidance as a result of using the satellite system is derived for the various ranges of interest. Over the range from 2 to 70 earth stations, the satellite system replaces terrestrial trunking facilities only, such that when $N = 70$, there is a satellite earth station at each backbone switch, eliminating essentially all terrestrial trunking charges. Over this range the following relationship appears to give a good fit to a number of simulation results that design the mixed media trunking arrangements:

$$C_{AN} = C_t(1 - e^{-N/15}),$$

where

$$C_t = m_t C_{tm} + 2n_t C_{tt} \qquad \text{and} \qquad 0 \leq N \leq 70$$

This exponential relationship drives the terrestrial trunking cost to nominally zero for $N = 70$ while achieving only a modest reduction in trunking costs for values of N less than 15. For example, for $N = 5$, $C_{AN} = \$128$ million, or about 25% of the total trunking costs. Looking at the actual situation, if five ground stations are employed in the United States, about half of the trunking mileage costs could be avoided, but no termination costs could be saved since lines would still be required from the switches to the satellite earth stations. It should be emphasized that a large fraction of the trunking mileage can be saved using a relatively small number of satellite earth stations because demand-assignment satellite systems have the unique property of complete connectivity as new stations are added. In other words, it is presumed that each satellite earth station can communicate directly with every other earth station, with no tandem switching of traffic to add capacity requirements to the trunking cross-sections. Each satellite earth station then provides for a reduction in terrestrial trunking from its vicinity to the vicinity of all other earth stations, simultaneously. As the number of earth stations increases above five, colocation of satellite earth stations with switches becomes practical, leading to savings in both mileage and termination costs. When $N = 70$, all switches are presumed to have colocated satellite terminals, leading to a cost avoidance equal to the full trunking cost.

Cost Avoidance with More Earth Stations

When N is greater than 70, the cost avoidance is equal to C_t (the total trunking cost) plus the access line cost to each user access area location that has a satellite earth station within it. The number and average length of the access lines for each size category of access area location can be determined by reference to the user profiles in Table 12-1.

As the number of earth stations increases from 70 to 210, earth stations are being deployed at the large user concentrations, each requiring an average of 87 access lines over a distance of 135 miles to the serving switch. Thus, in the range of $70 \leq N \leq 210$,

$$C_{AN} = C_t + 87\{(\$110)(135 \text{ miles}) + (\$13,750)\}(N - 70)$$
$$C_{AN} = C_t + 2.49 \times 10^6(N - 70) \quad \text{for } 70 \leq N \leq 210$$

When $N = 210$, the value of C_{AN} is computed as:

$$C_{AN} = 489 \times 10^6 + 2.49 \times 10^6(140) = \$837 \times 10^6$$

In the range of $210 \leq N \leq 560$, the medium-sized access areas are accommodated with satellite earth stations, each one avoiding 23 access lines with an average length of 70 miles. This leads to a value of C_{AN} of:

$$C_{AN} = \$837 \times 10^6 + 23\{(\$110)(70) + (\$13,750)\}(N - 210)$$
$$C_{AN} = \$837 \times 10^6 + 0.493 \times 10^6(N - 210) \quad \text{for } 210 \leq N \leq 560$$

When $N = 560$, the value of C_{AN} is computed as:

$$C_{AN} = \$837 \times 10^6 + \$173 \times 10^6 = \$1010 \times 10^6$$

Similarly, for $560 \leq N \leq 1050$, the small-sized access areas are accommodated with satellite earth stations, each one avoiding 6 access lines, with an average length of 90 miles. This leads to a value of C_{AN} of:

$$C_{AN} = \$1010 \times 10^6 + 6\{(\$110)(90) + (\$13,750)\}(N - 560)$$
$$C_{AN} = \$1010 \times 10^6 + 0.142 \times 10^6(N - 560) \quad \text{for } 560 \leq N \leq 1050$$

When $N = 1050$, the value of C_{AN} is computed as:

$$C_{AN} = \$1010 \times 10^6 + 70 \times 10^6 = \$1080 \times 10^6$$

By combining the various equations for C_S and C_{AN} over the different ranges of interest, we can write a set of overall cost equations from the basic expression:

$$C_N = C_0 + C_S - C_{AN}$$

Over the range $2 \leq N \leq 50$:

$$C_N = \$1.169 \times 10^9 + \$52 \times 10^6 + 1.65NC_l$$
$$- \$489 \times 10^6(1 - e^{-N/15})$$
$$C_N = \$732 \times 10^6 + \$489 \times 10^6(e^{-N/15}) + 1.65NC_l$$

Over the range $50 \leq N \leq 70$:

$$C_N = \$1.169 \times 10^9 + \$104 \times 10^6 + 1.65NC_m$$
$$- \$489 \times 10^6(1 - e^{-N/15})$$
$$C_N = \$784 \times 10^6 + \$489 \times 10^6(e^{-N/15}) + 1.65NC_m$$

Over the range $70 \leq N \leq 100$:

$$C_N = \$1.169 \times 10^9 + \$104 \times 10^6 + 1.65NC_m - \$489 \times 10^6$$
$$- \$2.49 \times 10^6(N - 70)$$
$$C_N = \$958 \times 10^6 + 1.65NC_m - \$2.49 \times 10^6(N)$$

Over the range $100 \leq N \leq 210$:

$$C_N = \$1.169 \times 10^9 + \$132 \times 10^6 + 1.65NC_m - \$489 \times 10^6$$
$$- \$2.49 \times 10^6(N - 70)$$
$$C_N = \$986 \times 10^6 + 1.65NC_m - \$2.49 \times 10^6(N)$$

Over the range $210 \leq N \leq 350$:

$$C_N = \$1.169 \times 10^9 + \$132 \times 10^6 + 1.65NC_m - \$837 \times 10^6$$
$$- 0.493 \times 10^6(N - 210)$$
$$C_N = \$567 \times 10^6 + 1.65NC_m - 0.493(N)$$

Over the range $350 \leq N \leq 560$:

$$C_N = \$1.169 \times 10^9 + \$198 \times 10^6 + 1.65NC_m - \$837 \times 10^6$$
$$- 0.493 \times 10^6(N - 210)$$
$$C_N = \$633 \times 10^6 + 1.65NC_m - 0.493(N)$$

And over the range $560 \leq N \leq 1050$:

$$C_N = \$1.169 \times 10^9 + \$198 \times 10^6 + 1.65(560)C_m + 1.65(N - 560)C_s$$
$$- \$1010 \times 10^6 - 0.142 \times 10^6(N - 560)$$
$$C_N = \$436 \times 10^6 + 924C_m + 1.65C_s(N - 560) - 0.142 \times 10^6(N)$$

These equations can be readily evaluated for various values of N, C_l, C_m, and C_s to determine the cost relationship as a function of the number of earth stations in the system. The results of such an evaluation are shown in Figure 12-4, where the total system transmission costs are plotted as a function of the number of earth stations, for many possible values of earth station costs.

The uppermost curve in Figure 12-4 uses very conservative estimates of C_l, C_m and C_s, with the large earth stations assumed to cost \$4.4 million each, the medium earth stations \$3.5 million each, and the small earth stations \$1.4 million each. At the other extreme, the lowest curve shown uses values of \$1.0 million, \$0.3 million, and \$0.1 million, respectively. At the present time, it is most likely that the actual earth station costs will be closer to the low curve plotted in this figure. Recent progress in satellite-radiated power and high-gain satellite antennas has materially reduced the need for cryogenically cooled receiver front ends. Improved satellite station keeping substantially reduces the need for tracking systems in the earth stations. Improved large-scale integration implementations of the TDMA/DA equipment in the earth stations has significantly reduced the channel bank and modulation equipment at each earth station. The mass produc-

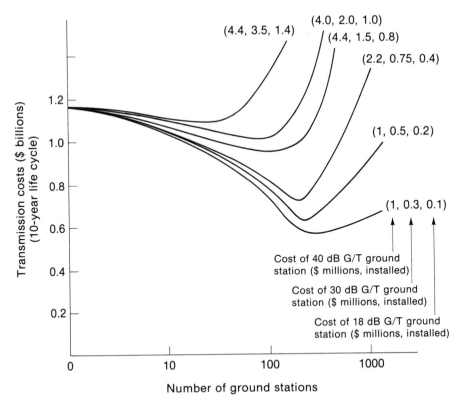

Figure 12-4.   Total system transmission costs as a function of number of earth stations.

tion of many elements as a result of the strong demand for television receive-only earth stations has reduced the overall station costs and made available a large body of technical and structural expertise in the siting, design, and construction of earth stations.

As we can see from Figure 12-4, the optimum value of the number of earth stations—that is, the number of earth stations that minimizes the total system cost—increases from about 75 earth stations for the highest assumed value of earth station cost, to a value of about 500 earth stations at the lowest set of costs. In order to show the relative importance of the various components of the system cost curves, Figure 12-5 redraws one of the curves from Figure 12-4 showing the components of the total cost. Note that any discontinuities between the various segments of the curves have been fitted and smoothed so as to present a single overall curve for each set of parameters, even though the generating equations have a number of discrete ranges of interest.

Figure 12-5 indicates that, by the time the system employs about 50 earth stations, the cost of the ground segment is more than that of the space segment. This leads to the observation that adding complexity and capability to the satellites would have a leveraged impact on overall system costs, given commensurate reductions in ground station costs. Improvements in satellite transmitter power,

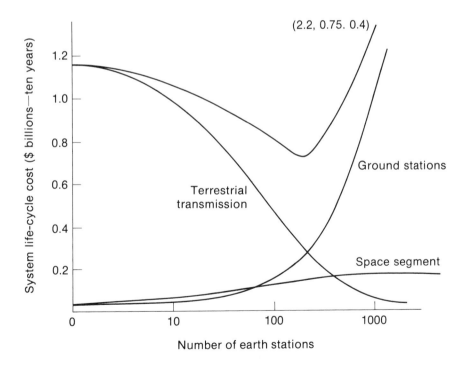

Figure 12-5.   Breakout of system costs.

through higher power or larger satellite antennas, appears to be the most direct way to improve the balance between the space and earth segment costs. However, other technical advances such as satellite-borne switching, advanced signal processing, forward error correction, adaptive antenna arrays, or the use of additional frequency bands and diversity techniques can all have the desired impact. If such a tradeoff is achieved, thereby reducing ground system costs through higher space segment costs, not only would the minimum of the total cost curve be reduced in absolute magnitude, but the location of the minimum would move toward a higher optimum number of ground stations.

Under the assumptions of the analysis, the majority of the earth stations used in the system are medium sized. The sensitivity of the results to the assumed cost of this size earth station is shown in Figure 12-6, where the optimum number of earth stations is plotted as a function of the cost per earth station for the medium-sized (30 dB G/T) station. Here again, we see that the optimum number of earth stations falls in the range of 100 to 300 for the model shown, with that number

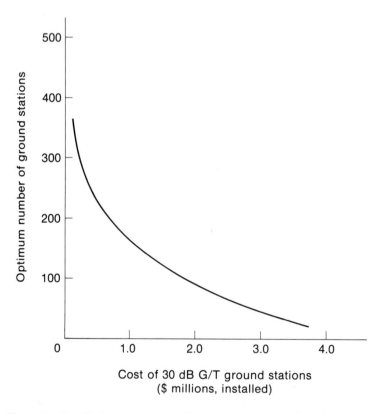

**Figure 12-6.   Optimum number of earth stations as a function of cost
per medium-sized station.**

increasing rapidly if the cost of the earth stations can be driven significantly below a half million dollars—a likely occurrence at the present rate of improvement in satellite technology.

FURTHER THOUGHTS ON COMBINED SATELLITE/TERRESTRIAL NETWORKS

Our analysis, approximate though it is, has shown that satellites can have a significant impact on development of new communications networks if a fairly large number of ground stations are employed. Network cost savings approaching 50% of the life-cycle cost compared to terrestrial services–only networks can be achieved. The optimization of network costs using satellites could evolve over a long period of time, driven by the technology and costs of the earth stations, with satellite services displacing terrestrial services as the economics permit.

We must emphasize that these results are derived using a simplified model that aggregates the projected traffic in the entire United States into 70 service areas, although a total of more than 1000 specific service locations were examined. The analysis accounts only for transmission cost savings, even though substantial reduction in switching costs are likely in any system that makes liberal use of demand-assigned satellite capacity. Many of the access techniques, resource sharing techniques, and capacity assignment techniques we have discussed throughout this book are directly applicable to the issue of reduced switching cost, through the satellite visibility and end-user terminal based traffic control and processing abilities.

Finally, it is important to remember that, although this analysis and model represent a rather large set of total requirements, using more than $100 million per year in transmission services, the analysis techniques and sensitivity to system parameters apply over a broad range of public and private networks. In addition, because of the United States "open skies" policy for domestic satellite systems, and the availability of full transponder bandwidths from many satellite vendors, implementation of this kind of a mixed satellite/terrestrial network is feasible for a broad range of user organizations, either individually or through shared systems or user group private systems. The techniques demonstrated in this case study can be applied in a broad range of systems and user applications to determine system feasibility and optimization of the various media. In the next chapter we will derive a very simple relationship by which the results of our analysis can be rapidly approximated by a single, simple equation.

SUMMARY

1. Large-scale telecommunications networks combining satellite and terrestrial connectivity can be designed by aggregating user requirements into a simplified model and deriving total network cost relationships.

2. The basic concept of network optimization attempts to use just enough satellite connectivity to offset the most costly terrestrial portions of the network.

3. In the model examined in this case study, employing approximately 2500 different network service points, the optimum design used between 100 and 400 earth stations, depending on earth station cost.

SUGGESTED READING

ROSNER, ROY D. "Optimization of the Number of Ground Stations in a Domestic Satellite System." *Proceedings of the EASCON Conference, EASCON '75 Record.* Washington, D.C., September–October 1975, pp. 64A–64F.

The first published study introducing the concept of satellite communications directly to large user concentrations, with satellite capacity used on an end user-to-end user basis, this paper is the basis for much of the material in this chapter. The published paper provides several additional considerations omitted for simplicity here, including power budgets in the satellite links, the demand assignment mechanism, the effect of quantity discounts in using large numbers of similar earth stations, and so on.

13

A Random Network Approach to Distributed Satellite Networks

THIS CHAPTER:

will reconsider the problem of optimizing the mix of satellite and terrestrial facilities in a large network, using a more general approach.

will derive minimizing criteria for system cost based on user density and overall system element costs.

will apply these criteria to several examples to illustrate their effectiveness and ease of use.

Satellites can be combined with any of many different resource sharing and access protocols to achieve flexible, responsive, cost-effective distributed network designs. Inherent in such approaches is the assumption that satellites form the essence of the distributed network and all users have immediate local access to the satellite earth stations. If a satellite earth station is not available locally, terrestrial connections must tie the individual users to the nearest earth station, or else a mixture of terrestrial switching nodes and satellite resources must be used. A first-order approach to introducing satellites to a distributed network, with or without distributed switching, is to use the satellites to tie the terrestrial nodal facilities together. The next step would logically be for the satellite facilities, in a broadcast mode employing ALOHA or any of the more complex and capable resource sharing techniques, to provide all the interswitch trunking among the network nodes. However, only when the satellites can provide the connections completely back to the individual users will the total benefit of the networkwide broadcast and distributed access control be realized.

It is possible to develop a general model of a highly satellite-based distributed network that is less detailed than the model in the last chapter. The "macro" model that will be derived here, though it will not tell us where to put the earth stations, will provide an initial estimate of the number of earth stations needed for an optimum distributed network design.

SERVING USERS WITHOUT DEDICATED EARTH STATIONS

The basic premise in this model is that the number of users, and the number of user locations, is sufficiently large to assume random geographical user distribution over the area being served by the system. This assumption is analogous to the random traffic generation rates, leading to the Poisson traffic arrivals, that we have used throughout the book in analyzing capacity–delay relationships for various access techniques.

We will assume initially that, although the users are randomly distributed, the satellite earth stations are arranged in a regular, geometric pattern (called a **tessellated pattern**). Later, we will generalize the situation to randomly distributed earth stations and compare the results to the tessellated case.

Given these assumptions, we are working with a fixed geographic area in which network users are uniformly but randomly distributed. At most, there could be one earth station associated with each user, but in general an earth station acts as a concentration point, serving a number of users in its vicinity. The optimization criterion will be least cost of the system.

In order to conceptually simplify the geometry, we will assume the service area is a square, each side of which is S miles. We will derive the results in terms of user densities; in other words, the actual shape of the service area is not important provided that the area is big enough to support a fairly large number of earth stations, and the average distance between earth stations is a small fraction (1/4 or less) of the smallest dimension of the service area. The assumed geometry is based on the square service area shown in Figure 13-1.

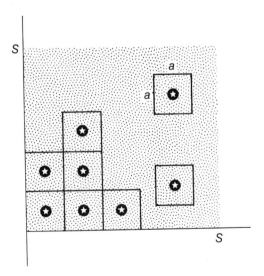

Figure 13-1.   Satellite network geometry for random users and tessellated servers. © 1976, ICCC.

The overall service area of S^2 square miles is assumed to be divided into q service areas. Each small service area, surrounding one of the earth stations, is a small square, with sides of dimension a. In order to completely cover the service area, we note that:

$$qa^2 = S^2$$

or

$$a = S/\sqrt{q}$$

Additional parameters needed for this model are:

q = number of earth stations

n = number of user locations

S = dimension of total service area

a = dimension of service area of a single earth station

p = density of users = n/S^2

C_q = average cost of an earth station

C_l = cost per mile of line needed to tie each user to the
nearest earth station

$\bar{d}$ = average distance from each user to the nearest earth station

The overall cost of the network comprises four major components: the earth station cost, the space segment cost, the terrestrial line cost, and the terrestrial fixed plant costs.

Earth Station Cost

The earth station cost, represented by C_q, is the average installed cost of the satellite earth stations. Since an economic analysis is involved, the earth station cost will represent the life-cycle cost of the station—including its acquisition cost, installation cost, cost of operation, maintenance, and repair over a ten-year life cycle.

Space Segment Cost

The space segment cost represents the overall cost of the satellite, including launch and continuing control costs. Since no single-purpose network is likely to use the full capacity of an entire satellite, this cost would be the pro-rated share of the total capacity allocated to the particular network of interest. In addition, since the amount of satellite capacity used is dependent on the amount of user traffic, and not the number of earth stations used to route that traffic through the satellite, we will assume that the space segment cost is constant and independent of the number of earth stations that ultimately optimize the network. The underlying implication is that all user traffic eventually reaches the satellite, via either a local or a remote earth station.

Terrestrial Line Costs

The terrestrial line costs are the mileage costs associated with the connection from each user to his nearest earth station. For n users, each an average of $\bar{d}$ miles from the nearest earth station, the total line costs are just: $n\bar{d}C_l$, where C_l is the cost per mile of the terrestrial lines. To be commensurate with the earth station costs, the terrestrial line costs would have to be estimated for a comparable life-cycle period, say ten years, based on lease tariff charges, which are generally given as monthly costs.

Terrestrial Fixed Plant Costs

The terrestrial fixed plant costs are the costs independent of distance and traffic—for instance, attachment charges, modem costs, port costs, or any other costs related to the number of users but not to the cost of the earth station itself. Since all users will eventually be connected to an earth station somewhere in the network, these costs are essentially independent of the number of earth stations.

Cost Minimization Equation

The problem, therefore, reduces to a cost minimization, thus:

Minimize total system cost, C, where:

$C =$ (earth station cost) $+$ (terrestrial line costs) $+$ (space segment cost) $+$ (terrestrial fixed plant costs)

$C = qC_q + nC_l\bar{d} +$ (space segment cost) $+$ (terrestrial fixed costs)

The independent variable here is the number of earth stations, q, and our intention is to find the number of earth stations that minimizes the overall cost. Since the space segment and terrestrial fixed plant costs are not dependent on the number of earth stations, but only on the number of users and the total amount of traffic flowing through the satellite, we need not consider them in the minimization process. The problem thus reduces to:

Minimize C, with respect to q, where:

$$C = qC_q + nC_l\bar{d}$$

The key to the minimization process is the recognition that the average distance between the users' locations and the nearest satellite earth station, $\bar{d}$, is directly related to the number of earth stations, q. If q is very large—that is, there are many earth stations—the average distance to the nearest station will be small. If the number of earth stations is small, and the earth stations are sparse, then the average distance to the nearest station will be larger. There will be less cost in the earth stations and more cost in the terrestrial lines connecting the users to the nearest earth stations.

DERIVING THE OPTIMUM NUMBER OF EARTH STATIONS

Analysis for Stations in a Tessellated Pattern

The general service area of the network, as was shown in Figure 13-1, consists of a random scattering of users, divided into small, square service areas, with a satellite earth station located in the center of each service area. The geometry within one of the small service areas is shown in Figure 13-2. The side of the square was defined as a, where the value of a is related to the total service area and the number of earth stations by the relationship $a = S/\sqrt{q}$. Inside the small service area are a sprinkling of user locations, some close to the earth station at the center of the square and some much further from the earth station.

The average distance from the earth station to the users is determined by a mathematical integration process, which in effect computes the distance to every possible user point within the square and takes the average of those distances. Let us look at the small shaded area in Figure 13-2, located a distance r from the center, and at an angle θ with respect to the horizontal axis. The area of this small shaded portion is $dr \times r\, d\theta$ since the linear dimensions are given in the figure by the incremental notation. All of the points within the tiny incremental area are essentially a distance r from the center, so that the weighted average of all of the distance within this incremental area is $r \times dr \times r\, d\theta$ divided by the total area

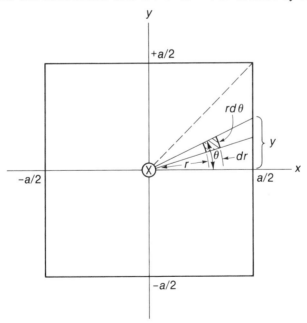

Figure 13-2.   Geometry of a user service area.
© 1976, ICCC.

of the service area, or $r^2 \, dr \, d\theta / a^2$. The overall average for the entire service area is then obtained by integration as:

$$\bar{d} = \int_0^{2\pi} \int_0^r r^2 \, dr \, d\theta / a^2$$

Performing the integration is generally straightforward, except that the upper boundary value of r depends on the value of θ and on the quadrant of the service area in which the integration is proceeding. By symmetry we can see that integrating over the entire square is only eight times the value of integrating over the first half of the first quadrant. Over this limited area, the maximum value of r can be found from:

$$\text{Cos } \theta = \frac{a/2}{r}$$

or

$$\text{Sec } \theta = \frac{r}{a/2}$$

Making these substitutions:

$$\bar{d} = \frac{8}{a^2} \int_0^{\pi/4} \int_0^{a/2 \, \sec \theta} r^2 \, dr \, d\theta$$

Evaluation of this integral is not overly difficult with the aid of a table of integrals; the elaboration can be found in Abramson and Rosner (1976). The result is simply:

$$\bar{d} = 0.381a$$

which says that the average distance from the center of a square to every other possible point within the square is 38% of the side of the square. We can take this value back to the cost formulation, where we find that:

$$C = qC_q + nC_l\bar{d} = qC_q + nC_l(0.381a)$$

or, since $a = S/\sqrt{q}$:

$$C = qC_q + 0.381nC_lS/\sqrt{q}$$

This gives us the expression fully relating the overall cost of the satellite-based network as a function of the number of earth stations. We can now proceed to optimize this cost by taking derivatives of the expression and setting it equal to zero. Thus:

$$\frac{dC}{dq} = 0 = C_q - 0.381nC_lS/2(q)^{3/2}$$

Solving for q, we find that:

$$q_{\text{opt}} = \left[\frac{0.381nC_lS}{2C_q} \right]^{2/3}$$

By remembering that the user density, p, is given by: $p = n/S^2$, we can generalize the optimization by substituting for S, with the result:

$$q_{\text{opt}} = 0.332n \left[\frac{C_l}{C_q p^{1/2}} \right]^{2/3}$$

This expression relates the number of users, the user density, the cost of the lines connecting the users to the earth stations, and the cost of the earth stations. From the expression we find that the optimum number of earth stations increases if there are more users and if the cost of the lines connecting users to earth stations is high. The optimum number of earth stations is lowered by high earth station costs and high user densities because many users can then be served by fewer earth stations.

Analysis for Stations Located Randomly

If the earth stations are located randomly rather than in a tessellated pattern, a different mathematical approach is required in order to estimate the optimum number. Figure 13-3 shows the situation of a user location surrounded by number of randomly distributed earth stations. Within the total service area there are q possible earth stations that can serve the user, but we desire to compute the

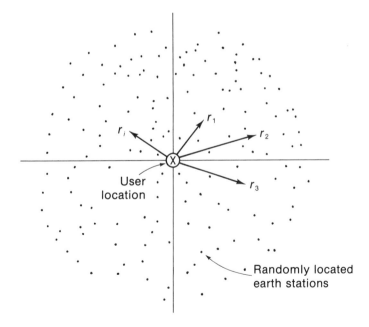

Figure 13-3. Network geometry with random users and random earth stations.
© **1976, ICCC.**

average distance to the nearest possible earth station that can serve each user. There is, in effect, a double averaging over two random processes. First we must compute, for each user, the average distance to the nearest earth station. Then we must compute the overall average over a total user population within the service area. Once again, the details are elaborated in Abramson and Rosner (1976); we will simply indicate the results for the random earth station case:

$$q_{\text{opt}} = 0.396n \left[\frac{C_l}{C_q p^{1/2}} \right]^{2/3}$$

This result differs from the previous result only in the scale factor, showing that the optimum number of earth stations for the random case is about 20% higher than when earth stations are arranged in a regular, geometric pattern. The effect is due primarily to the fact that, when the earth stations are randomly distributed along with the users, there is a probability that the random clusters of earth stations may be away from the clusters of users, leaving some earth stations much less loaded than others. When the users are random but the earth stations are distributed regularly, there will be, on the average, a much more balanced usage of the earth stations, leading to a lower optimum number.

Tessellated Earth Stations

$$q_{\text{opt}} = 0.332n \left[\frac{C_l}{\sqrt{p}C_q} \right]^{2/3}$$

Random Earth Stations

$$q_{\text{opt}} = 0.396n \left[\frac{C_l}{\sqrt{p}C_q} \right]^{2/3}$$

q_{opt} = optimum number of satellite earth stations

n = number of network users

C_l = cost of the terrestrial connection from the users to the nearest earth station (per mile–life cycle)

C_q = cost of the satellite earth station (life-cycle cost)

p = user density = n/A, where A is the total area being served by the network

Box 13-1

For convenience, both of these results are summarized in Box 13-1; we will demonstrate their application by examples in the next section. Remember, the intention and application of this approach to the satellite/terrestrial network leads to an estimate of the average number of earth stations needed to optimize the network configuration. The actual location of the earth stations would depend on the particular topology of the user situation.

APPLICATIONS AND EXAMPLES OF THE RANDOM NETWORK ANALYSIS

A Network of Many Small Users

The application of the formulations in Box 13-1 can best be illustrated by several examples. In the first example we will consider a network of many small users, distributed throughout an area the size of the continental United States.

The applicable parameters are:

n = number of users = 30,000

S^2 = area of the continental U.S. = 3,000,000 square miles

p = user density = 30,000/3,000,000 = 0.01 user per square mile

C_l = line cost = \$2.00 per mile per month
This monthly cost can be equivalenced to an approximate ten-year, present-worth cost by multiplying the annual cost by a ten-year, 10% present-worth factor of about 6.5. Thus, the ten-year present worth of a \$2.00-per month expenditure is approximately $2 \times 12 \times 6.5 = \$156$.

C_l = \$156 per mile for ten years

C_q = \$250,000 = ten-year installed and operated cost of the satellite earth stations

By substituting these parameters in the equation for the optimum number of earth stations in the regular geometric pattern, we find:

$$q_{opt} = 0.332 \times 30,000 \left[\frac{156}{(250,000) \times (0.01)} \right]^{2/3}$$

or

$$q_{opt} = 9960 \times (6.24 \times 10^{-3})^{2/3}$$
$$q_{opt} = 9960 \times (3.37 \times 10^{-2})$$
$$= 336 \text{ earth stations}$$

Using the formulation for random, rather than tessellated, distribution of earth stations results in an optimum number of earth stations of 401 compared to

336. This result is interesting since a network of 30,000 users, though large, is not uncommon for many shared network applications, such as point of sale networks, transactional networks, banking terminals, and many other such examples. In addition, it is not at all difficult to visualize a ten-year cost of about $250,000 for a modest earth station using present state-of-the-art technology.

Earth Stations Serving Clusters of Users

Our second example will consider a situation where the earth stations are intended to serve clusters of users rather than individual users. This case demonstrates that it is possible to treat user clusters in the same way as individual users. In addition, this case will provide results that we can compare directly to the results of the detailed optimization analysis in Chapter 12.

Following Table 12-1 (Chapter 12), we will assume a total of 2500 user locations within the area of the continental U.S., with potential network users clustered at these locations. After doing a local traffic analysis on the "average" location, we determine that each location needs ten lines to connect the installation to the nearest satellite earth station. The applicable parameters for this case are:

n = number of users = 2500

S^2 = area of the U.S. = 3,000,000 square miles

p = user density = 2500/3,000,000 = 0.000833 user per square mile

C_l = line cost = $2.00 per line per mile, per month, times an average of ten lines per user location = $1560 per mile for the ten-year evaluation period

C_q = $1,500,000 for the ten-year period for a relatively large earth station capable of serving many users

By substituting these parameters into the equation for the optimum number of earth stations using the regular geometric pattern, we find:

$$q_{opt} = 0.332 \times 2500 \left[\frac{1560}{(1,500,000) \times \sqrt{0.000833}} \right]^{2/3}$$

or

$$q_{opt} = 830(36.1 \times 10^{-3})^{2/3}$$
$$q_{opt} = 830(11.0 \times 10^{-2}) = 91 \text{ earth stations}$$

Using the formulation for random, rather than tessellated, distribution of earth stations results in an optimum number of 109 earth stations. The number of earth stations is considerably smaller for this case because the number of individual user locations is smaller, and the assumed cost of the earth stations is much higher than in the first example. In fact, the optimum number of earth stations is so highly dependent on the cost of the stations that it is generally

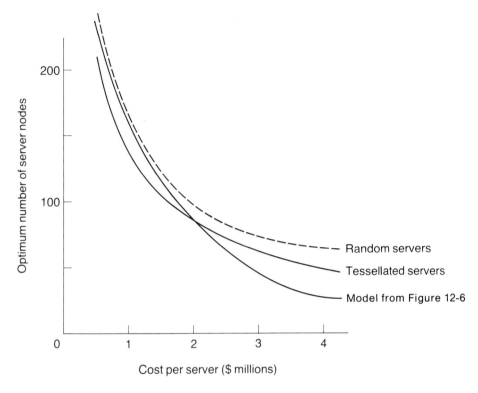

**Figure 13-4.   Optimum number of earth stations plotted against
cost per station. © 1976, ICCC.**

useful to plot the optimum number of earth stations as a function of their assumed
life-cycle cost. This is done for this set of data and plotted in Figure 13-4, where
the optimum number of earth stations is plotted over a range of earth station
costs between zero and about $4,000,000. Both the random and tessellated arrange-
ments are plotted, along with the curve from Figure 12-6 (Chapter 12) for comparison. The agreement is amazingly good between the random analysis curves
and the curve from Figure 12-6. Notice that the divergence between the curves
increases at small values of the optimum number of earth stations, since the
averaging assumptions of a random network become poor if there are too few
earth stations.

USING THE TOOLS OF MACROANALYSIS

We have used two considerably different approaches to illustrate distributed
network design. Detailed network design, using actual or estimated user locations,
may appear necessary at first, but useful results can be rapidly achieved with

randomization and averaging techniques. Early phases of network feasibility design and overall structuring of the network can be carried out with the techniques that have been demonstrated throughout this book. Of course, eventually the detailed design and specific location of each network facility will have to be determined. Despite the many sophisticated algorithms that have been developed, specific network designs will generally be determined by the available locations of the network implementer and the commercial facilities that will provide the bulk of the transmission. In other words, detail design ultimately ensues from the macroanalysis combined with the pragmatic situation.

SUMMARY

1. Large telecommunications networks can be approximated by random distributions of user and service facilities in order to determine the initial design parameters of a distributed, mixed media network.

2. The specific approach taken randomized the users, and then considered earth stations placed in regular (tessellated) and random patterns.

3. The resulting equations from the analysis solved for the optimum number of earth stations as a function of system parameters and costs, including user density, terrestrial line costs, earth station cost, and total number of users.

4. The approximate results were found to correlate well with results from much more detailed analyses.

SUGGESTED READING

ABRAMSON, NORMAN, and ROSNER, ROY D. "Optimum Densities of Small Earth Stations for a Satellite Data Network." *Proceedings of the International Conference on Computer Communications, ICCC '76.* Toronto, Canada, July 1976, pp. 123–127.

This paper provides the basic analysis of the random network approach to earth station optimization. The results are similar to those presented in this chapter, and the mathematical details of the optimizing equations are included.

14

The Communications Environment
in the Information Age

THIS CHAPTER:

will introduce the available common carrier and commercial
facilities from which distributed, integrated networks can be
synthesized.

will illustrate the market and tariff structures under which
those services are provided.

Throughout this book, we have tried to provide a practical understanding of
the concepts of modern advanced communications technology. We have empha-
sized those ideas and techniques that, when integrated, will permit the establishment of highly flexible, efficient, and reliable communications facilities for
organizations advancing further into the "information age."

In this chapter we will consider the combination of facilities necessary for
deployment of operational networks serving the needs of major communications
users. The balance of user-owned services and those provided by commercial
providers and common carriers depends on many factors, including capital
investment costs, individual requirements, geographical distribution of users, and
the availability of commercial services. In general, telecommunications users will
lease the long-haul transmission media portions of their networks, while owning
and operating some or all of the nodal and processing facilities. If satellite networking techniques are used, the number, deployment, and ownership of satellite
earth stations are also considerations.

COMMON CARRIER AND COMMERCIAL SERVICES

Most organizations likely to implement distributed communications networks
will probably acquire and own nodal devices, processors, multiplexors, concentrators, switches, and possibly satellite earth stations; very few are likely to own
the major interstate, intercity, and international transmission facilities. At the

same time, over the last decade the common carrier industry has shifted from a highly regulated, single-server system to a highly competitive industry, relatively responsive to improving technology and changing user requirements. Technological evolution to predominantly digital networks has also made the public networks more adaptable to a wider variety of user needs.

Common Carrier Competition

Early Trends. Competition in communications services has developed quite rapidly since the landmark **Carterfone decision** of 1968 and the approval of Microwave Communications Incorporated (**MCI**) construction of interstate microwave transmission facilities in 1969. These two rulings by the U.S. Federal Communications Commission (**FCC**) in effect declared that it was not in the public interest to restrict all telecommunications to the regulated monopolies of the established (voice) phone companies.

At the time, most experts expected the impact of these decisions to be competition in the area of terminals and end instruments. The construction and operation of long-haul transmission facilities is an extremely capital-intensive business, with very high start-up costs and strong economies of scale. The traditional microeconomic picture of a firm or industry supply curve rising to the right (that is, marginal cost per unit increasing with increasing quantity as less efficient resources are brought into the marketplace) does not apply to telecommunications. Modern technology has provided the ability to add large total capacities at large total costs but small unit costs, providing the demand is sufficient to justify the construction.

Despite the fact that the technology heavily favors the large, established carriers, many suppliers have moved into the telecommunications transmission marketplace and are competing successfully with the established phone companies. Their success is due partly to some protective regulation and fully allocated pricing demanded by the FCC, and partly to the ability of the new suppliers to apply the most modern satellite and computer-based techniques to their services. Furthermore, new entrants in the communications market can choose to enter the most lucrative service areas first—those where potential customers are most densely concentrated and can be served with large-capacity, high-efficiency facilities.

The Breakup of AT&T. The most dramatic changes in the competitive carrier marketplace are likely to take place over the next few years, as the 100-year-old structure of the world's largest corporation will be fundamentally altered. In 1980, the FCC, part of the executive branch of the government, ruled, through its Computer II Inquiry, that the American Telephone and Telegraph Company could begin to enter nonregulated markets so long as it created an "arms-length" subsidiary. At the same time, the FCC established the concept of dominant and nondominant carriers, which made it easier for carriers to offer new and expanded services. Dominant carriers are defined under the FCC rules as those service providers that can exercise monopolistic market power within their service areas. Nondominant carriers, which included most of the non-AT&T companies, could implement a new service or modify their rates simply by filing the appropriate tariffs with the FCC; they would not have to wait for FCC approval.

At the same time, the Department of Justice was proceeding with an antitrust suit against AT&T, based on allegations that AT&T acted anticompetitively in its dealings with the competitive carriers through the early and mid-1970s. This suit led to a proposed negotiated settlement in early 1982; if finally approved, the settlement will lead to the breakup of the Bell System into two, disjoint companies—regulated local operating companies, on the one hand, and the nationwide interstate carrier network on the other. Such separation of the Bell System will presumably permit equal access by all interstate carriers to the local telephone plant, while also permitting the interstate carriers to compete with each other more readily. In addition, the proposed settlement would permit AT&T to enter other markets, freeing the company from the restrictions of the 1956 decree that kept AT&T from the data processing and information services field.

Meanwhile, the U.S. Congress has, since 1976, been trying to develop a satisfactory replacement for the Communications Act of 1934. Early versions of the rewrite were a result of such massive lobbying efforts by the Bell System that drafts of the act were known as the "Bell Bill." Recent versions attempt to balance the needs of the public, the needs of national security and emergency preparedness, and the needs of the highly technical industry.

Regardless of the specific Communications Act, the basic situation is clear: the competition of commercial communications services will be permanent, and the potential user of those services will have to choose carefully among them to be sure of maximum efficiencies and economies. In addition, competition will enhance the speed with which new technology reaches the marketplace and the speed with which transmission/communications services are blended with information services, local networks, and user data processing devices to increase productivity and move completely into the information age.

The Market and Tariff Structure

Utilizing the currently available carrier services, particularly as the transmission medium between service points of distributed communications networks, requires an understanding of the various elements of the cost structure of those services. Carrier facilities can be broadly classified according to speed, system arrangement, and rate structure. Figure 14-1 illustrates the interrelationship of these characteristics. The lines connecting the choices of speed, system arrangement, and rate structure form a logical decision tree of possible service combinations. Note that packet switching or satellite services do not enter into the basic classification of services; they are actually different media capable of supplying services under this structure.

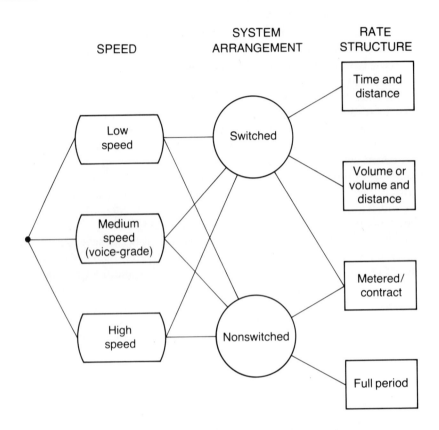

Figure 14-1. Interrelationship of characteristics of common carrier services.

Speed. The speed (or bandwidth) characteristic recognizes the use of a voice channel as the fundamental building block of telecommunications services. Low speed refers to rates and bandwidths typically 300 bits/second or less, many of which can fit into a single voice-grade channel. Medium speed (or voice-grade) describes the capacity of a single voice-equivalent channel. Such a channel is required for voice-to-voice conversation, but, with present technology, it is also capable of providing 9600 bits/second of data transmission. High speed (or wideband) applies to all capacity requirements exceeding a single voice-equivalent channel, all the way up to the hundreds of megabits of data needed for video and sensor data. By and large, normally tariffed common carrier services go up to 56,000 bits/second, with higher rates generally needing special agreements and special arrangements, often available only via satellite facilities.

System Arrangement. The system arrangement characteristic divides telecommunications services into switched and private-line (nonswitched) services. In a switched network service, there are fewer lines than users, in the expectation that not all the users will need to communicate at the same time. Users gain not only economy but also the flexibility of being able to communicate with a large number of other users by indicating to the switches the address (destination) of the other party.

Private-line systems use dedicated, point-to-point lines between the communications endpoints. The line is always available, but is fixed in connectivity, such that only the users connected to that line are able to use the point-to-point capacity. When the dedicated users are idle, the capacity is wasted. It is possible to lease point-to-point circuits on a dedicated basis and use customer-owned switches, processors, or controllers to effectively establish a switched network at the user level. This is the essence of what have come to be known as value-added networks (**VANs**), which are based on one organization leasing point-to-point circuits from a major carrier and embedding them in a sophisticated switched network. As far as the primary carrier is concerned, the leased circuit is permanently terminated at the customer endpoints, even if those endpoints happen to be switches. The lower path through Figure 14-1 is thus very important in the development of private networks since the full-period circuits provide the major connectivity of those systems.

Rate Structure.　　The variety of rate structures makes it difficult to compare the services of the various common carriers. Transmission rate, distance, data quantity, connection time, total communications load, time of day, geographical locations, and a host of other data all must be taken into account. For simplicity, four broad rate structure classifications have been indicated in Figure 14-1.

Full-period service, applicable only to private-line systems, permits 24-hour-per-day, 365-day-per-year usage of the line. Private-line services are also provided on a metered, or monthly contract minimum usage, basis. The implication of the latter is that, although the user thinks he has a dedicated private line, the carrier is somehow able to use that line (or portions of it) for other customers and is thus willing to offer a lower rate than for full-period service. Metered and monthly minimum contract services are also available using the switched networks, the most common example being the Wide Area Telephone Service (**WATS**) for nationwide long-distance ("toll-free") calling.

Many rate structures are sensitive to time, distance, and possibly time of day. The best example is the direct-distance dialing (**DDD**) service of the phone company, where the cost of the call depends on its length (in minutes), the distance between the end offices completing the call, and the time of day and day of the week the call takes place.

Finally, services tailored primarily to data communications have established rate structures based primarily on volume, or a combination of volume and distance. This approach is especially applicable to packet switching services where usage charges are based on the number of packets or data characters actually transmitted. Note that volume is not synonymous with time, since volume of data involves a combination of usage time, data rate, and idle periods. The value-added and satellite services have been tending toward rate structures of this type.

Another form of tariff structure is of increasing importance in the context of private networks, especially those utilizing satellite facilities. Because certain types of demand vary greatly between daytime and nighttime, and because satellite carriers generally have to maintain part of their capacity in the form of unused space facilities aboard on-orbit satellites, it is sometimes possible to lease satellite capacity on a "preemptable" basis. Such capacity, which can be reclaimed by the vendor under prescribed conditions, can often be obtained at a rate far below the standard tariff for similar capacity on a permanent basis.

WORLDWIDE COMMUNICATIONS SERVICE PROVIDERS

The common carrier marketplace is considerably different in North America than in most of the rest of the world. In most countries, telecommunications services are provided by the national postal, telegraph, and telephone (PTT) administrations of each country as monopoly activities. In fact, in most countries, the rate structures of the telecommunications services are specifically designed to subsidize the operations of the national postal system, with communications services priced above cost and postal services priced below cost. International services are generally set through negotiated agreement among the countries, acting in pairs or through various consortiums.

Suppliers of POTS

In the United States, the common carrier market consists of more than 1500 telephone and telecommunications companies, most organized on a corporate, profit-making basis and a few being cooperatives. Most of these companies are very small, regionalized independent phone companies supplying **POTS** ("plain old telephone service") service to a town, region, or community. Approximately 82% of the telephones in the United States are served directly by the American Telephone and Telegraph Company (AT&T) through one of its 21 operating companies. AT&T also owns the Western Electric Manufacturing Company and Bell Laboratories, which do the research, development, and product manufacturing for the entire Bell system. AT&T supplies most of the nation's interstate long-haul communications through the AT&T Long Lines Department. About 15% of the nation's phones are served by the General Telephone and Electronics Corporation (GT&E), also through local and regional operating companies. GT&E owns Lenkurt Manufacturing Company (transmission equipment) and Automatic Electric Company (switching equipment). GTE-Telenet provides packet switching network services, and GT&E Research Laboratories provides research and development for the corporation.

The remaining phones in the United States are served by other local and regional companies, some as small as rural offices serving only 50 subscriber lines. The key point is that, within each of their areas of jurisdiction or franchise, each of these companies acts as a regulated monopoly supplier of POTS service. All of these companies use compatible technical standards, procedures, and parameters to insure total interoperability and interconnectivity of the telephone equipment anywhere in the total system.

Specialized Services

Competition has arisen and flourished in the so-called specialized services market. Specialized services is a catch-all phrase used to describe any and all communications services (voice and data) that do not require total interconnectability with the rest of the world's telephones. Principal offerings originally stemmed from data communications services designed to serve computer-based applications, but rapid expansion of the specialized market has included voice, video, graphics, facsimile, teletype, message, and other communications functions.

Many local and regional specialized carriers operate within the United States. Tables 14-1 through 14-4 provide a brief sketch of the services offered by the organizations that are capable of providing nationwide services. It should be

Table 14-1. Packet Switching Data Communications Carriers

GTE-TELENET COMMUNICATIONS CORPORATION

Nationwide value-added, switched service network

First commercial packet switcher

Distance-independent services—volume-dependent costs

Gateway services to non-U.S. networks

Electronic mail services

Virtual private networks

TYMNET, INCORPORATED

Nationwide value-added, switched service network

Evolved from Tymshare time-sharing ADP services

Distance-independent services—volume-dependent costs

Direct and gateway services to non-U.S. locations

Electronic mail services

Virtual private networks

GRAPHNET COMMUNICATIONS

Nationwide value-added, switched service network

Provides graphics and facsimile services

Limited gateway services to non-U.S. locations

Provides data format processing and conversion

Table 14-2. Domestic Satellite Carriers

AMERICAN SATELLITE CORPORATION

Joint venture of Fairchild Industries and Continental Telephone

First U.S. domestic satellite carrier

Uses capacity leased from various satellite spacecraft

First to use customer-premises earth stations

COMSAT GENERAL CORPORATION

U.S. domestic subsidiary of international COMSAT Corporation

Presently a "carrier's carrier"—capacity leased to GT&E, AT&T

Owns and operates three COMSTAR satellites

Provides direct services via one-third partnership in SBS

RCA AMERICAN COMMUNICATIONS CORPORATION

Major distributor of cable TV services to customer premises
(cable TV operator–owned) earth stations

Private-line services via shared and customer earth stations

Wideband (56 Kb/s) data point-to-point services

WESTERN UNION CORPORATION

Customer-premises earth stations or microwave interconnect

Private-line voiceband and wideband digital services

Leased transponder as well as leased channel services

SATELLITE BUSINESS SYSTEMS (SBS)

Partnership of IBM, COMSAT General, and Aetna Insurance

Integrated digital voice, data, facsimile service

Customer-premises and shared earth stations

PROPOSED NEW SYSTEMS

Hughes Communications Corporation

Southern Pacific Communications Corporation

General Telephone and Electronics Corporation (GSAT)

American Telephone and Telegraph Corporation

Table 14-3. Domestic Terrestrial Carriers

MICROWAVE COMMUNICATIONS INCORPORATED

First authorized specialized interstate common carrier
Nationwide private-line service
Switched voiceband services—EXECUNET
Extensive private terrestrial microwave network
Primarily uses phone company services for local access

SOUTHERN PACIFIC COMMUNICATIONS CORPORATION

Microwave carrier using railroad right-of-way
Extended to nationwide service with acquisition of DATRAN assets
Nationwide private-line services
Switched voiceband services—SPRINT
Uses phone company services for local access

ITT DOMESTIC TRANSMISSION SYSTEMS, INCORPORATED

ITT entry into domestic private-line and switched voice services
Evolved from ITT acquisition of United States Transmission Services
Switched voiceband services—CITY-CALL
Domestic value-added services—data, facsimile, and message
FAXPAK and electronic mail services
Gateway services to non-U.S. locations

PROPOSED DIGITAL TERMINATION SERVICES (DTS)

Satellite Business Systems (SBS)
Tymnet
Local Data Distribution Corporation (LDD)
GTE-Telenet
Western Union
MCI
ISACOMM
Graphic Scanning Corporation (Graphnet)
DTS Corporation
National Microwave Interconnect
Data Services Corporation
Via/Net Corporation
Contemporary Communications Corporation
RCA

217

Table 14-4. Overseas Gateway Carriers

ITT WORLD COMMUNICATIONS, INCORPORATED

Major gateway carrier between domestic services of many countries

Provides telex, telegram, leased channel, and message services

Universal Data Transfer Service provides packetized gateway
services between Telenet, Tymnet, and foreign networks

RCA GLOBAL COMMUNICATIONS CORPORATION

Major international carrier among 200 locations in U.S. and overseas

Provides telex, telegram, leased channel, and message services

WESTERN UNION INTERNATIONAL, INCORPORATED

Major gateway carrier between domestic services of many countries

Provides telex, telegram, leased channel, and message services

Handles satellite maritime, television, and voice/data services

WUI Database Service provides packetized gateway data
facilities among U.S. and about 20 foreign countries

Provides high-speed data services via satellite

TRT TELECOMMUNICATIONS CORPORATION

International carrier between United States, Caribbean, Latin America

Provides telex, telegram, and message services

Produces and distributes STORTEX—high-speed telex system

COMMUNICATIONS SATELLITE CORPORATION (COMSAT)

Furnishes satellite services to common carriers providing satellite
service between U.S. and foreign countries

Owns 23% interest in International Telecommunications Satellite
Organization (INTELSAT)

Operates major research laboratory devoted to satellite technology

noted that, in all cases, AT&T, either directly or through its operating companies, can offer a competitive service to any service supplied by the specialized carriers, with the exception, at the present time, of direct point-to-point satellite transmission and packetized value-added services. Furthermore, a number of technically advanced services are under development by various organizations, including the Advanced Communications Service (ACS) being developed by AT&T.

Another attribute of the specialized common carriers is that they can limit their services to those areas where the market is large enough to support the entry of new carriers. Because not all carriers operate in all cities and states, the applicability of many services depends on the carrier's area of coverage. In order to use some of these services, users commonly lease a point-to-point circuit from AT&T to carry the connection from the user's location to the nearest service point of a particular carrier. Even if the specialized carrier is tariffed directly to the city where the customer has his facilities, the initial connection from the customer's premises to the specialized carrier's offices often has to be supplied by the local phone company. This last problem may be overcome by the recent FCC authorization for the development of digital termination services (**DTS**) in many metropolitan areas. DTS will use microwave, point-to-point radio systems, operating in the 10.5 GHz band, to interconnect user locations with specialized carrier nodal locations. Most major metropolitan areas currently have as many as nine carriers with approved applications to develop DTS capabilities.

The ultimate combination of common carrier facilities and intelligent user end devices to form fully distributed communications networks is still a complex mixture of art and science, judgment and instinct. The rapid changes brought about by technology and competition for equipment and services quickly turns today's system obsolete. Specific system and tariff costs have to be updated frequently, and a number of commercial services do this regularly (see Center for Communications Management, and Datapro Research Corp., 1978).

Many organizations, because they do not fully appreciate the opportunities and challenges of distributed communications networks, are overpaying for their communications and information distribution services. We have explored, throughout this book, many approaches that can help a small telecommunications staff in any organization establish a baseline and set of alternative configurations for reducing costs and improving services. Because of the resources generally involved in information processing and movement in any modern organization, distributed communications networks usually produce very handsome returns on investment.

SUMMARY

1. The multitude of specialized common carriers, each with its own features, service areas, and attributes, and the rapidly evolving technology, regulatory environment, and competition create a confusing picture of the best way to meet telecommunications needs.

2. A variety of telecommunications facilities, both leased and tariffed, are available. They can be classified according to speed, system arrangement, and rate structure.

3. Among the common carriers in the United States, POTS is supplied by regulated monopoly companies, primarily AT&T, although that picture is likely to change in the middle 1980s. By contrast, specialized services are provided by many different companies offering a great variety of applications.

SUGGESTED READING

Center for Communications Management, Inc. "Executive Telecommunications Planning Guide," updated periodically. Available from the Center for Communications Management, Ramsey, NJ 07446.

The Center for Communications Management is one of the consultant publication services that provide extracts of current communications tariffs for ready use by communications planners. The guide is in a loose-leaf format, and the publisher provides replacement pages when tariff changes are made.

Datapro Research Corp. "All About Data Communications Facilities." Datapro Research Report No., 70-G-100-01a, May 1978. Available from Datapro, Delran, NJ 08875.

This report compares and contrasts the various data communications carriers and their services. It presents various tariffs available in the industry, ways to optimize the use of services, expected changes in services, and information about the various termination arrangements and their charges. Check with the publisher for more recent versions of this report.

SHAW, LOUISE C. "Data Communications Carriers." *Datamation*, vol. 26, no. 8 (August 1980), pp. 107–112.

This article highlights the companies that provide data communications services, primarily in the United States or at gateways between the United States and various other countries. The article also provides key financial data.

TAYLOR, CAROL A., and WILLIAMS, GERALD. "Considering the Alternatives to AT&T." *Data Communications*, vol. 9, no. 3 (March 1980), pp. 47–62.

This comparative description of Tymnet, Telenet, American Satellite, and Western Union Satellite services describes the overall technology utilized by each of the carriers along with information on the service areas, and basic tariffs and charges for each service.

GLOSSARY

ACS. *A*dvanced *C*ommunications *S*ervice. A new common user data communications service developed by AT&T. Recently renamed Advanced Information Service/Net I.

ALOHA. The broadcast multiple access technique that permits users to transmit packets whenever they desire.

Analog channel. A communications channel that responds linearly to changes in the frequency and amplitude of the information transmitted. Will accurately represent the input signal over a specified range of parameters.

APC. *A*daptive *P*redictive *C*oding. A voice conversion technique that uses information about the human voice to achieve very low bit rates for normal speech inputs.

ARPANET. The computer network developed by and for ARPA (the Advanced Research Projects Agency within the U.S. Department of Defense). It has been the basis for much of the technology of packet switching.

Baseband. A local networking technique that directly modulates the direct current data bits onto the transmission medium, at data rates typically as high as 10 megabits/second.

Bit. Contraction of *BI*nary *D*igit. A single symbol, either a one or a zero, which, when used in groups, represents the numbers, letters, and other symbols of communications. Generally used in groups of 5, 8, or 16.

Blocking. A phenomenon in a communications network where one user cannot reach another because of any one, or a combination of, network resource limitations.

BPS. *B*its *P*er *S*econd; sometimes written as B/S or b/s. A measure of the speed with which data communications can move over a line. The prefixes K (for thousand) or M (for million) are often added to represent higher speeds.

Broadband. A local networking technique that translates the direct current digital signal to a higher modulation or operating frequency before placing it on the transmission medium. This permits frequency-division multiplexing of many high-speed data signals onto a single system, achieving total data rates of hundreds of megabits/second.

Broadcasting. Communications achieved by transmitting the information from a central point over a shared medium, so that it can be heard by all potential users at the same time.

Buffer. Part of a communications processor or switch used to store information temporarily.

Bursty traffic. Communications traffic characterized by short periods of high intensity separated by fairly long intervals of little or no utilization.

Bus. A facility for connecting a number of processors to a common point or common connection medium (such as a cable), via which data can be directly exchanged on a processor-to-processor basis.

Call. A complete, two-way interchange of information between two or more parties in a network, extended over a period of time. It will generally consist of a number of sequential messages or transactions passed over the communications circuits in each direction.

Call request. The initiation of a new call into a network, preceded by information that tells the end user addresses needed to establish the call.

Capture. A communications phenomenon whereby the stronger of two signals captures the receiver and remains relatively insensitive to interference from the weaker signal.

Carterfone decision. A landmark decision in federal communications regulation that permitted the attachment of foreign devices—those not supplied by the franchised carrier—to the telephone network. It opened the way for competition in the communications marketplace.

CCITT. *C*onsultative *C*ommittee for *I*nternational *T*elephone and *T*elegraph. An international advisory committee set up under United Nations sponsorship to recommend standards for international communications.

Channel. A single physical communications medium capable of moving intelligence from one point to another. Specific physical and electrical parameters generally define its capacity.

Ciphertext. The output of encoding systems that is a changed version of the input text and cannot be read or understood without going through a decoding operation.

Circuit switching. A form of switched network that provides an end-to-end path between user endpoints under the control of the network switches. Often called channel switching.

Collision. The condition where two packets are transmitted sufficiently close in time that some portion of each intersects and interferes with the other.

Common carrier. An organization, generally franchised by a governmental body, to provide communications services to the general public.

Concentrator. A device that improves the efficiency of a communications circuit by fitting a number of low-speed inputs into a single, higher speed, output that has a lower speed than the sum of the individual input speeds.

Congestion. A network condition that causes information to be delayed or interrupted, even though capacity may be available elsewhere in the network.

Contention. A transmission facility sharing technique that allows devices to transmit at will into a commonly available channel or medium. A variety of different protocols may be used to resolve interference (contention) resulting from (nearly) simultaneous transmission.

Continuous traffic. Communications traffic that (nearly) completely fills the resource while it is in progress; for instance, the transmission of a television program. Compare **bursty**.

CRT. Literally, *C*athode *R*ay *T*ube. Used in a generic sense to refer to data terminals that display transmitted and received data on a televisionlike screen.

CSMA. *C*arrier *S*ense *M*ultiple *A*ccess. A method of contention operation whereby the terminals sense the state of the channel before attempting transmission.

CVSD. *C*ontinuously *V*ariable *S*lope *D*elta Modulation. A method for converting analog speech into a digital format by transmitting a signal that is proportional only to the difference between two successive samples of the original analog signal.

Data encryption standard. A federal standard issued by the U.S. National Bureau of Standards in 1977 that provides a methodology for encoding and decoding digital information.

Datagram. A mode of packet network operation whereby the contents of a single packet are handled as a distinct entity with no functional connection with the preceding or following packets.

DDD. *D*irect *D*istance *D*ial Network. The major nationwide long-distance network, provided mainly by the Bell System, that permits users to complete connections to distant users without the assistance of an operator.

Decryption. The mathematical/electrical process that decodes **ciphertext** back to a readable or understandable format.

Delay. The additional time introduced by a communications network in delivering a specified amount of data compared to the time that the same information would take on a full-period, point-to-point circuit.

Delta modulation (DM). A simplified form of **CVSD** that converts analog speech into a digital format by transmitting a signal proportional to the difference between two successive samples of the original analog signal, and which always uses the same step size between samples.

DES. see **Data encryption standard.**

Downlink. The transmission path from a satellite to the satellite earth station to which it is transmitting.

DPCM. *D*ifferential *P*ulse *C*ode *M*odulation. A method that permits the conversion of analog information into a digital format by encoding the difference in amplitude between successive samples.

DTS. *D*igital *T*ermination *S*ystem. A common carrier service recently authorized by the U.S. FCC that uses a microwave signal to directly connect a central station in a metropolitan area to the user end location.

Duplex channel. A communications channel capable of transmitting information in both directions at the same time.

Encrypted communications link. A communications circuit that is protected from intrusion by putting the information through an encoding process before it is transmitted over the link.

Encryption. The process of encoding information before transmission to prevent its interception or intrusion.

Erlang. A measure of communications traffic intensity representing the full-time use of a communications facility. For example, one erlang represents the traffic that can be carried on a single line used continuously for one hour.

Ethernet. A form of contention operation, being commercially deployed by Xerox Corporation, used to tie facilities together in a local geographic area.

Facsimile transmission. The transmission of still images over a communications facility by encoding the black-and-white portions of the original image with different digital codes.

FCC. *F*ederal *C*ommunications *C*ommission. The principal regulatory body in the United States responsible for interstate communications and common carrier services.

FDM. *F*requency-*D*ivision *M*ultiplexing. A means whereby a number of separate communications circuits are combined over a common facility.

FDM/FM. A technique, often used in microwave and satellite transmission, whereby an FDM signal, a composite of many individual channels, is transmitted at a single radio frequency using frequency modulation.

Gateway. A node or switch that permits communication between two dissimilar networks.

Group. A number of communications channels handled as a single entity.

Half-duplex channel. A communications circuit capable of carrying traffic in either direction, but only one way at a time.

Header. The initial part of a data block or packet that provides basic information about the handling of the rest of the block.

Hold and forward. A switching technique where each message is held at intermediary switches long enough for the accuracy of the received information to be checked before it is relayed on to the next switch.

Image communications. Similar to facsimile transmission but may also include the slowly varying video information.

Intelligent terminal. A data communications terminal that has sufficient intelligence (processing power) to perform fairly complex interface functions and local formatting and processing.

Interarrival time. A statistical measure of the average time between successive new calls or messages to a network.

Internetwork gateway. A node or switch that permits communications between two dissimilar networks.

ISDN. *I*ntegrated *S*ervices *D*igital *N*etwork. A single network capable of equally handling voice, data, image, and other traffic forms, all in digital format, over common facilities.

ISO. *I*nternational *S*tandards *O*rganization. An international body that standardizes goods and services. ISO works in conjunction with CCITT for standards that impact communications.

Leader. The initial part of a user data block that tells the network the destination and handling of the following data.

Link. A physical or electrical connection between two endpoints; for communications purposes may consist of one or more channels.

Logical channel number. A designator of an apparent connection via a packet switched network, by time sharing the channel to the network switch.

Logical multiplexing. The ability of a given user to communicate with various network destinations simultaneously, using a single access line, by time sharing the line and using different logical channel numbers for each connection in progress.

Loop. A local connection of user devices forming a closed path, providing two paths (either direction around the loop) by which information can reach other users connected to the loop. Often used interchangeably with *rings*.

LPC. *L*inear *P*redictive *C*oding. A technique for converting analog speech into digital format; uses information about the human voice to synthesize the transmitted speech with a very small number of transmitted bits.

LSI. *Large Scale Integration.* The implementation of modern electronic circuitry where hundreds or thousands of electronic components are integrated onto a single silicon chip about one-half inch square.

MCI. *Microwave Communications, Incorporated.* One of the first common carriers licensed to compete with the Bell System for interstate communications services.

M/D/1 queue. Generalized notation of **queueing** theory, designating a delay process characterized by Poisson (M) arrivals, deterministic (D) message lengths, and a single (1) server.

Mesh. With respect to local networks, an interconnection of various elements on a point-to-point basis, with little centralized management or overall structure.

Message switching. Switching technique where messages are stored in their entirety at each intervening switch.

M/M/1 queue. Generalized notation of **queueing** theory, designating a delay process characterized by Poisson (M) arrivals, Poisson (M) distributed message lengths, and a single (1) server.

Modem. *MOdulator-DEModulator.* A device that allows digital signals to be transmitted over analog facilities.

MPL. *Multischedule Private Line.* A tariff of the Bell System that provides for full-period, point-to-point analog circuits between service locations within the United States.

Multidrop. The data communications analogy of a party line, where various user terminals are connected on a common shared line.

Multiplexing. A communications technique that combines a number of low-speed channels into a single higher speed channel.

Narrowcasting. The logical inverse of broadcasting, where any one of many different locations may transmit information in toward a central point. Also applied to the situation where a central transmission point "broadcasts" information that can be heard by many receivers, but the information is selectively addressed to only one, or a few of the many, possible receivers.

NCC. *Network Control Center.* The major control point of a network; collects performance data and issues control commands.

NCP. *Network Control Program.* The software that must be executed by a front-end computer in order to interface with a packet switched network in the full packet mode.

Node. A point of a network where various links come together; generally contains a switching element used to direct traffic.

Overhead. Information required by a network for its operation over and above the basic information that is being moved on behalf of the subscribers.

Packet switching. A network technique that divides the user messages into relatively short blocks (packets) and uses numerous geographically distributed switching nodes to achieve low end-to-end delay for real-time traffic.

PCM. *Pulse Code Modulation.* A technique for coding analog signals for transmission on a digital circuit, by sampling the analog signal at regular intervals and converting each sample into a digital codeword.

Persistence.   Characteristic of a user who continuously monitors the occupancy of a channel and transmits a packet as soon as he detects that the channel is idle.

Poisson arrival process.   A specific probabilistic description of the number of new messages or call arrivals into a communications system over a given time interval.

Polling.   A technique that permits a large number of terminals to share a common channel. A central controller asks each terminal, in turn, to transmit any information it may currently have queued.

POTS.   *P*lain *O*ld *T*elephone *S*ervice. The common user voice-based nationwide telephone network.

Protocol.   A set of rules and procedures that permit the orderly exchange of information within and across a network.

PVC.   *P*ermanent *V*irtual *C*ircuit. A logical connection across a packet network that is always in place and available; used to emulate a full-period connection.

Queueing.   Any process that combines elements of storage and delay with a number of servers. The delay experienced by the users of the process can be estimated on the basis of statistical behavior of the various elements involved.

Reservation technique.   Any of a number of possible packet broadcast methods requiring that users reserve capacity in advance of transmitting their data.

Ring.   A local connection of user devices forming a closed path, providing two paths (either direction around the ring) by which information can reach other users connected by the ring. Often used interchangeably with *loop*.

RJE.   *R*emote *J*ob *E*ntry. A data terminal used to enter complete jobs for processing at a remote computer location.

Robust.   A network characteristic indicating the ability to operate nearly normally when certain network elements fail.

Routing.   The process of finding a suitable path to move information through the network.

SBS.   *S*atellite *B*usiness *S*ystems. A partnership of IBM, COMSAT, and Aetna Insurance companies to provide private network services via satellite communications facilities.

Segment.   A part of an overall information exchange that is transmitted between the user device and the network. It may be the same length as or longer than a packet, depending on the protocol implementation.

Sisyphus distance.   A theoretical distance in a radio-based network, at which a station is sufficiently far from the central receiver that it is unable to successfully deliver traffic to the central point because of interference of closer (and therefore stronger) stations.

Slotted ALOHA channel.   A packet transmission time that has a fixed (rather than random) relationship to allowable transmission times of other packets.

SPADE.   Single Channel per Carrier Demand Assignment. A method of transmitting digitized signals by assigning each one to a separate frequency channel.

Spoofing.   The use of a network to transmit data that claims to be coming from a source other than its actual source.

Star.   A local network configuration where each user device is tied directly back to a single centralized controller.

Store and forward. Switching technique where each message is stored in full at each switch it passes through.

SVC. *S*witched *V*irtual *C*ircuit. A logical connection across a packet switched network. It is established on an as-needed basis and can provide connection to any other switched user in the network.

Synchronous. A form of communications where characters or bits are sent in a continuous stream, with the beginning of one contiguous with the end of the preceding one. Separation of one from the other requires the receiver to maintain synchronism to a master timing signal.

Synchronous loop. A local network configuration with the processing elements placed in tandem with the elements of the loop, and using synchronous transmission from element to element.

TADI. *T*ime *A*ssignment *D*igital *I*nterpolation. A digital technique for interleaving data bursts during silent intervals in voice conversations.

Tandem switches. Switches in a network that provide a path between other switches, rather than originating or terminating traffic.

Tariffs. The formalized charges for telecommunications services that are filed and approved by state and federal regulatory organizations.

TASI. *T*ime *A*ssignment *S*peech *I*nterpolation. A technique for carrying a group of voice channels over a physical facility by interleaving conversations in the idle periods of other voice conversations.

TDM. *T*ime-*D*ivision *M*ultiplexing. A means whereby a number of separate communications circuits are combined over a common facility by dividing the common facility into discrete time intervals.

TDMA/DA. *T*ime-*D*ivision *M*ultiple *A*ccess/*D*emand *A*ssignment. A sharing technique for satellite capacity that uses time-division multiplexing but assigns the time slots on the usage demand of individual users, rather than on a permanent basis.

Tessellated pattern. A physical arrangement of facilities (such as nodes or earth stations) that follows a regular, geometric pattern.

Timeout period. The length of time that a switch or processor will wait for an expected action (such as an acknowledgement) before it takes unilateral action.

Token. The message or sequence of bits that, when received at a network device, gives that device permission to transmit its information on the common communications medium. Generally used in loop or ring networks, and frequently referred to as *transmit tokens* or *write tokens*.

Topology. The physical arrangement of facilities, links, and nodes, to form a network, including the connectivity pattern of the network elements.

Transaction. A computer-based message that represents a complete unidirectional transfer of information between two points on a data network.

Transparent. An operation or set of operations that occur in such a way that the user of the system is unaware any such activities are taking place.

Transparent switching. A network switching technique that handles information by providing a logical or physical end-to-end connection.

Trapdoor function. A mathematical function or operation that is much easier to perform in one direction than in the inverse direction. For example, it is much easier to find the square of a number than to find its square root.

Trojan horse threats. Threats to network security by the insertion of programming steps into a software program. The inserted steps, which are difficult to detect other than by the original programmer, can provide private information.

Trunk. The communications circuit between two network nodes or switches.

TVRO. *TeleVision Receive Only.* A satellite earth station designed to receive the downlink signals of a satellite solely for the purpose of distributing television programming.

UDN. *Ultimate Distributed Network.* A hypothetical network that provides direct user access to a nationwide network with small portable terminals, and which can communicate with a variety of modes and facilities.

Uplink. The transmission path from a satellite earth station to the satellite itself.

User lockout. A phenomenon on shared circuits where a user is temporarily inhibited from transmitting because of temporary overload of the available circuits.

VAN. *Value-Added Network.* The class of public network that leases facilities in the form of basic transmission from one carrier, adds intelligence, and provides more "valuable" services to end users.

VFCT. *Voice-Frequency Carrier Telegraph.* A technique that permits the combination of up to 24 teletype channels over a single voice-frequency channel.

Video transmission. The transmission of a moving image, such as a television picture, by electronic means.

Virtual circuit. A logical connection across a packet switched network that emulates a point-to-point circuit by insuring data integrity, transparency, and data sequence.

WATS. *Wide Area Telephone Service.* A nationwide long-distance phone service, where users contract for high-volume circuit usage rather than paying for each call individually.

X.25. The international standard developed by **CCITT** that provides the foundation for public packet switching networks.

Index